Space Technology Library

Volume 41

Editor-in-Chief
James R. Wertz, Microcosm, Inc., El Segundo, CA, USA

The *Space Technology Library* is a series of high-level research books treating a variety of important issues related to space missions. A wide range of space-related topics is covered starting from mission analysis and design, through a description of spacecraft structure to spacecraft attitude determination and control. A number of excellent volumes in the *Space Technology Library* were provided through the US Air Force Academy's Space Technology Series. The quality of the book series is guaranteed through the efforts of its managing editor and well-respected editorial board. Books in the *Space Technology Library* are sponsored by ESA, NASA and the United States Department of Defense.

Bobby Nejad

Introduction to Satellite Ground Segment Systems Engineering

Principles and Operational Aspects

 Springer

Bobby Nejad, Ph.D.
European Space Agency
Noordwijk, The Netherlands
Bobby.Nejad@esa.int

ISSN 0924-4263 ISSN 2542-8896 (electronic)
Space Technology Library
ISBN 978-3-031-15902-2 ISBN 978-3-031-15900-8 (eBook)
https://doi.org/10.1007/978-3-031-15900-8

This Springer imprint is published by the registered company Springer Nature Switzerland AG
The registered company address is: Gewerbestrasse 11, 6330 Cham, Switzerland

*To my daughter Julia, may she find her way
to reach the stars.*

Foreword by David H. Atkinson

The greater our knowledge increases,
the more our ignorance unfolds.
John F. Kennedy (1962)

The current epoch of spacecraft exploration of the planets began on December 14 in 1962, when the Mariner 2 spacecraft flew past Venus, making only 42 minutes of observations. Since then, many more interplanetary spacecraft have left Earth, including flybys, orbiters, entry probes, and landers to explore Mercury, Venus, our moon, Mars, Jupiter, Saturn, the planet-sized moon Titan, Pluto, and even across the known boundaries of our Solar System. The two Viking orbiter and lander spacecraft arrived at Mars in 1976, the Pioneer and Voyager spacecraft encountered the giant planets, and the Galileo orbiter with its probe provided the first *in situ* and long-term orbital explorations of Jupiter starting in 1995.

On 1 July 2004, Cassini entered orbit around Saturn and shortly thereafter released the Huygens probe to enter Titan's atmosphere and successfully parachute to the surface. To date, Huygens constitutes the furthest human outpost in our Solar System. The Cassini orbiter continued its mission to explore Titan's surface remotely and unveiled previously unseen mountains, valleys, dunes, seas, and shores. Today there are rovers on Mars and plans for extended spacecraft observations of asteroids and the moons of Jupiter, and planning for an aerobot mission to fly in the cold, dense atmosphere of Titan is underway.

From the first observations of the cosmos by the ancients to the start of telescopic observations of the skies in the seventeenth century to the current era of spacecraft exploration of the planets and heavens, the human imagination has been captured, answering the call to one of our most primitive desires - to discover and explore. A myriad of new environments in our solar system beckons, including possible subsurface oceans of Europa and other large moons of Jupiter, the volcanoes of Io, the geysers of Saturn's moon Iapetus and Neptune's moon Triton, and the Hellish surface of Venus.

Although the value of naked eye and telescopic observations cannot be overstated, the enormity of our Solar System requires sophisticated and increasingly complex vehicles designed to survive and properly function after many years in the treacherous environment of space, and for probes and landers to survive the extreme environments of other worlds.

Whether exploring the distant universe or remaining closer to home to study the Earth from orbit, all spacecraft require continuous support from ground operators to remain healthy and capable of performing the tasks for which they were designed. The maintenance of spacecraft in flight is only possible with the support of a highly complex and reliable ground infrastructure that provides operators the ability to reliably monitor, command, and control them, and give them the means to detect anomalies early enough to be able to take corrective actions and avoid any degradation or even loss of the mission.

In his book *Introduction to Ground Segment Systems Engineering*, Dr. Bobby Nejad presents a concise, complete, and easily understandable overview of the main architecture, working principles, and operational aspects of that essential ground infrastructure. In this treatise, Dr. Nejad covers the design, development, and deployment of ground systems that are vital to safely and efficiently operate modern spacecraft, be it Earth satellites, planetary orbiters, entry probes, landers, or manned space vehicles that all support humankind's continued quest to discover.

NASA Jet Propulsion Laboratory   David H. Atkinson, Ph.D.
Pasadena, CA, USA
November 2021

Foreword by Sonia Toribio

Whenever a brand new satellite has made it to the launch pad, ready for liftoff, all the attention first focuses on the impressive launch and later shifts to the first images that are transmitted from space. Not much thought is ever given to the flawless functioning of the complex ground infrastructure that had to be put in place beforehand, in order to ensure that operators can monitor and command the space segment throughout all the phases of the mission with high reliability, 24 hours a day, and 7 days a week.

Ground segment engineering is not a new domain at all and exists since human's early space endeavors. The advent of new key technologies in recent years, however, clearly reshaped many areas of ground segment systems engineering. Novel technologies at the IT level have reached a level of maturity that makes them a viable option for space engineering, which by its nature is a rather traditional and conservative field. Hardware virtualization as one example has clearly revolutionized ground segment design as it allows a significant reduction of the physical footprint through a more optimized use of resources and provides new ways to implement redundancy concepts in a seamless way for the operator.

On the other hand, new challenges have immersed. One of them is the need for ground operations scalability in view of the mega-constellations comprising several hundreds of space assets that have been placed into orbit recently. Another challenge is to deal with the growing frequency and threat of cyberattacks and the enormous damage they already cause today. This clearly puts a very high responsibility on the ground segment designer to put a cyber-resilient system into place.

Ground segment engineering is a highly multidisciplinary domain and demands expertise from a number of different technical fields. Hardware and software development, database management, physics, mathematics, radio frequency, mechanical engineering, network design and deployment, interface engineering, cybersecurity, and server room technology, to name only a few of them. With this in mind, the timing of the publication of *Introduction to Satellite Ground Segment Systems Engineering* is highly welcome, as it tries to cover a broad range of these topics in a very understandable, consistent, and up-to-date manner. It allows both beginners and experts to gain new insights into a very exciting field which so far is only scarcely addressed by existing literature.

European Space Agency   Sonia Toribio
Noordwijk, The Netherlands
October 2021

Preface

A review of today's space-related engineering literature reveals a strong emphasis on satellite and subsystem design, whereas only little is said on the ground segment design, development, deployment, and how it is operated. One explanation could be that given the harsh launch and space environment, the design, development, and qualification of a satellite are being considered the more challenging and demanding part of a space project. Also the project failure risk is mainly driven by the space segment, as there is only limited or usually even no direct human access to perform any in-orbit repair or maintenance activity.[1] The perception of the ground segment in contrast is often simplified to a set of computer servers that only assist the operators in their daily work like orbit calculations, steering of the TT&C communication antennas, or relaying of telecommands (TC) and telemetry (TM). The complexity of ground segment system design, assembly, integration, and verification might be considered as much lower compared to the corresponding activities for the space segment.

This imbalanced view shall be challenged here. One driving factor for this is the fact that the number of space missions that comprise not only one single satellite but a formation or a constellation of spacecraft has increased significantly in the last decades. These constellations often have to provide a highly reliable as well as continuous (24/7) service to the end user. Examples are navigation, telecommunication, television, or Earth observation (weather and remote sensing) satellite systems. To make this possible, new operational concepts had to be adapted which incorporate to a much larger extent automated satellite monitoring and control. In this case, the operator's role is shifted from a continuous interacting nature to a supervising one. Automated ground support reduces the risk of human errors and the operational cost, which benefits both the operating agency and the end user.

---

[1] The Hubble Space Telescope servicing missions between 1993 and 2009 with astronauts fixing or replacing items in-orbit should be considered as an exceptional case here, as the high cost and complexity of such an undertaking would in most cases not justify the benefit.

This development has a clear impact on the ground segment design. Its various subsystems (or elements) need to become more robust and able to deal with (at least minor) irregularities without human help. The amount and type of communication between the various elements need to increase significantly to allow the various subsystems to "know" more about each other and cooperate in a coordinated manner like an artificial neural network. This clearly requires more effort for the proper definition of interfaces to ensure that the right balance is kept: highly complex interfaces require more work in their design, implementation, and verification phases, whereas too simplistic ones might prohibit the exchange of relevant information and favor misunderstandings and wrong decisions.

Another important aspect is that the system design should never lose the connection to the operators who are human beings. This puts requirements on operability like a clear and intuitive design of the Man Machine Interfaces (MMI), and the ability to keep the system sufficiently configurable in order to support unforeseen contingency situations or assist in anomaly investigations.

No other technological field has developed at such a rapid pace as the computer branch. Processor speed, memory size, and storage capacities which were simply inconceivable just a decade ago are standard today, and this process is far from slowing down. But there are also entirely new technologies entering the stage, like the application of quantum mechanical concepts in computers with the quantum bit replacing the well-known binary bit, or the implementation of algorithms able to use artificial intelligence and machine learning when solving complex problems. Despite the fact that these new concepts are still in their infancy, they have already shown a tremendous benefit in their application.

For the ground segment designer, this fast development can be seen as blessing and curse at the same time. Blessing, as the new computer abilities, allows more complex software to be developed and operated or more data to be stored and archived. At the same time, fewer resources like temperature-regulated server room space are needed to operate the entire system. Curse, as the fast development of new hard- and software, makes it more difficult to maintain a constant and stable configuration for a longer time period, as the transition to newer software (e.g., a newer version of an operating system) or hardware (e.g., a new server model on the market) is always a risk for the ongoing operational activities. Well-tested and qualified application software might not work equally smooth on a newer version of an operating system, but keeping the same OS version might only work for as long as it is being supported by the hardware available on the market and software updates to fix vulnerabilities are available. Therefore, the ground segment designer has to consider obsolescence upgrades already in the early design phase. One elegant approach is to use hardware virtualization, which is explained in more detail in this book.

Today's ground segment design has only little in common with their early predecessors like the ones used for the Gemini or Apollo missions. What remains however unchanged is that efficient and reliable satellite operations can only be achieved with a well-designed ground control segment. To achieve this, ground segment design requires a high level of operational understanding from the involved systems engineers. Also vice versa, one can state that satellite operators can only develop sound operational concepts if they have a profound understanding of the ground segment architecture with all its capabilities and limitations. The different types of expertise required in the ground segment development phases are not always easy to gather in one spot. Very often the system design and assembly, integration, and verification (AIV) are performed by a different group of engineers as the one that is in the end operating the system. Most of the time, only limited interaction between these teams takes place which can lead to operator-driven change requests late in the project life cycle with significant cost and schedule impact. This book aims to close this gap and to promote the interdisciplinary understanding and learning in this exciting field. With such an ambitious objective in mind, it is difficult to cover all important subject matters to the level of detail the reader might seek. This book should therefore be considered as an introductory text that aims to provide the reader a first basic overview with adequate guidance to more specific literature for advanced study.

<table>
<tr><td>Galileo Control Centre</td><td>Bobby Nejad, Ph.D.</td></tr>
</table>

Galileo Control Centre   Bobby Nejad, Ph.D.
Wessling, Germany
6 September 2021

About this Book

This book targets two groups of readers equally, newcomers to space projects aiming to gain a first insight into systems engineering and its application in space projects, and more experienced engineers from space agencies or industry working in small, medium, and larger size projects who seek to deepen their knowledge on the ground segment architecture and functionality.

Despite the book's focus on ground segment infrastructure, this text can also serve readers who primarily work in the satellite design and development domain, considering the growing tendency to reduce ground-based spacecraft operations through an increase in onboard autonomy. A deeper understanding of the ground segment tasks and its overall architecture will allow an easier transfer of its functionality to the satellite's onboard computer.

The book will also serve project managers who need to closely interact with systems engineers in order to generate more realistic schedules, monitor progress, assess risks, and understand the need for changes, all typical activities of a ground segment procurement process.

Being written with the thought of an introductory textbook in mind, university students are also encouraged to consult this book when looking for a source to learn about systems engineering and its practical application, but also to simply gain a first insight into space projects, one of the most demanding but also exciting domains of engineering.

Contents

About the Author

Dr. Bobby Nejad is a ground segment systems engineer working at the European Space Agency (ESA), where he has been involved in the design, development, and qualification of the ground control segment of the European global navigation satellite system *Galileo* for the last 14 years. He received his master's degree in technical physics from the *Technische Universität Graz* and in astronomy from the *Karl-Franzens Universität* in Graz. He conducted his academic research stays at the *Université Joseph Fourier* in Grenoble, the *Bureau des longitudes* of the Paris Observatory, and the *Department of Electrical & Computer Engineering* of the University of Idaho. Bobby started at ESA as a young graduate trainee in 2001 to work on the joined NASA and ESA Cassini-Huygens mission to explore Saturn and its moons in the outer planetary system. Being a member of the ESA collocated team at the NASA *Jet Propulsion Laboratory*, he earned his Ph.D. on the reconstruction of the entry, descent, and landing trajectory of the Huygens probe on Titan in 2005. In the following year, Bobby joined the operational flight dynamics team of the German Space Operations Centre (GSOC), where he was responsible for the maneuver planning and commanding of the synthetic aperture radar satellites TerraSAR-X, TanDEM-X, and the SAR-Lupe constellation, the latter one being the first military reconnaissance satellite system of the German armed forces. Bobby joined the Galileo project at the European Space Technology and Research Centre of ESA in 2008 and has since then contributed to the definition and development of the ground segment's flight dynamics, mission planning, and TT&C ground station systems. Since 2012, he works at the Galileo Control Centre, where he supervises the assembly, integration, and qualification of the Galileo ground control segment.

Chapter 1
Introduction

The successful realisation of a space mission belongs to one of the most challenging achievements of human beings. Even if space projects are historically seen very young[1] they still have many things in common with some of the oldest engineering projects of mankind like the erection of the pyramids in Egypt, the design and production of the first ships or cars, or the construction of aqueducts, bridges or tunnels, to name a few examples. All engineering projects have a defined start and end time, a clear objective and expected outcome or deliverable, very project specific technical challenges to overcome, and limited resources in time, funding, and people. To avoid that technical challenges and limited resources jeopardise the successful realisation of a project, it has soon been understood that certain practises help to achieve this goal better than others. All the engineering processes and methodologies used to enable the realisation of a successful system can be summarised as *systems engineering* (SE). In every technical project one can therefore find a systems engineer who closely cooperates with the project manager and must be familiar with the relevant SE processes and their correct application. A dedicated study by the International Council of Systems Engineering (INCOSE) could proof that cost and schedule overrun as well as their variances[2] in a project decrease with an increase of SE effort [1]. To provide a basic introduction into the main SE principles and their application in space related projects is therefore one of the main objectives of this book (see Chap. 2).

Following the typical systems engineering approach to structure a complex system into smaller parts called subsystems, components, or elements, a space project is typically organised into the space segment, the ground segment, the launch segment, and the operations segment as shown in Fig. 1.1. This structure

---

[1] The first artificial satellite Sputnik has only been put into orbit in the year 1957.

[2] The variance of cost and schedule overrun refers to the accuracy of their prediction.

B. Nejad, *Introduction to Satellite Ground Segment Systems Engineering*, Space Technology Library 41, https://doi.org/10.1007/978-3-031-15900-8_1

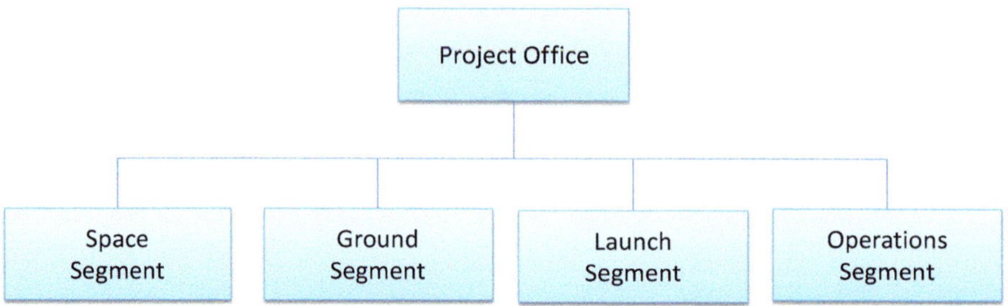

Fig. 1.1 High-level structure of a space project

is visible throughout the entire project life cycle and influences many aspects of the project management. It is reflected in the structure of the organisational units of the involved companies, the distribution of human resources, the allocation of financial resources, the development and consolidation of requirements and their subsequent verification, to only name a few.

At the very top sits the project office (PO) being responsible for the overall project management and systems engineering processes. The PO also deals with all the interactions and reporting activities to the project costumer who defines the high level mission objectives and in most cases also provides the funding. The customer can be either a single entity like a commercial company or in some cases a user community (e.g., researchers, general public or military users) represented by a space or defence agency. The PO needs to have a clear and precise understanding of the mission objectives, requested deliverables or services, and the required system performance and limitations (time and cost). These must be defined and documented as mission requirements and the PO is responsible for their fulfilment.

The launch segment comprises the actual launcher that has to lift the satellite into its predefined target orbit and the launch centre infrastructure required to operate the launcher during all its phases. The launch infrastructure comprises the launcher control centre, the launch pad, and the assembly and fuelling infrastructure. The organisational entity responsible for all the technical and commercial aspects of the launch segment is often referred to as the *launch service provider* (LSP). The LSP has to manage the correct assembly and integration of the launcher and the proper operations from its lift-off from the launch pad up to the separation of the satellite from the last rocket stage. The LSP must also ensure the launch safety for both the launch site environment (including the intentional self destruction in case of a catastrophic launch failure) and the satellite (e.g., disposal of the upper stage to prohibit collision with the satellite during its lifetime.)

The term *space segment* (SS) comprises any asset that is actually being put into space and requires its remote operation using the ground segment. It is obvious that there must be a well defined interface between these two segments which is called the ground to space interface. This interface must undergo a crucial verification and validation[3] campaign referred to as *system compatibility test campaign* (SCTC). This campaign has to be completed with the satellite still on ground, usually located at the satellite manufacturer's premises called the satellite assembly and integration test or AIT site. During the SCTC a set of flight operations procedures are being exercised and an actual exchange of telecommands (TCs) and telemetry (TM) needs to take place. Any error in this interface which is not discovered and fixed prior to the launch of the satellite can ultimately have a major impact in the operability of the satellite and could even lead to a loss of the entire space asset having a major impact on the project success. The satellite is a complex system in itself that consists of a number of subsystems. In most cases the satellite manufacturing and all the related project management is split into the payload part and the platform part. The payload is the part that is needed to perform the various mission tasks (e.g., remote sensing activities, broadcast of a TV, radio, or navigation signal, etc.) whereas the platform is everything else that is needed to guarantee the survival, proper functioning and operation of the payload. This separation and distinguished treatment of platform and payload is in some cases also supported by the fact that these two assets are being built by different companies and only integrated at a later stage in the satellite development life cycle. The satellite architecture and the specification of its subsystems will have an impact on how it has to be operated from ground. As the Ground Segment is required for satellite operations, its design will therefore also be impacted. Good ground segment engineering will benefit from a profound understanding of the satellite design and a brief overview of the common subsystems of a typical satellite platform is therefore presented in Chap. 3.

The ground segment comprises all systems installed on ground which are needed to operate the satellite platform and its payload. It should be noted that the term ground *control* segment or GCS is also used in this book and should be considered to have the same meaning.[4] The most relevant subsystems (also referred to as elements) are described in dedicated chapters which aim to provide an overview of the main functionalities but also the required inputs, expected outputs, and typical performance parameters. Each element itself consists of a hardware part (e.g., racks, severs, storage devices, workstations, monitors, etc.) and the software running on it (i.e., usually the operating system with a set of application software). Commonalities in the hardware equipment will always be a preference for any ground segment design as this will bring advantages for the maintenance of the system (e.g., common spare parts and suppliers, similar maintenance procedures for the replacement of failed parts, similar requirements for the rack design and deployment, etc.). The same will usually not be true for the application software

---

[3] The exact definition of the terms verification and validation and its difference is explained in Chap. 2.

[4] In certain projects the ground segment might be divided into smaller segments with different responsibilities, e.g., one part to operate the satellite platform and another one specialised for payload operations. This is usually the case when payload operations is quite involved and requires specialised infrastructure, expertise, or planning work (e.g., coordination of observation time on scientific satellite payloads).

part. Due to the different domain specialities the various application software might differ significantly in their architecture, complexity, operability, and maintainability. Therefore these might often be developed, tested, and deployed by different entities or companies which are highly specialised in very specific domains. As an example, a company might be specialised on mission control systems but has no expertise on flight dynamics or mission planning software. This diversity will put more effort on the system integration and verification part and the team responsible for this activity.

The integration success will strongly depend on the existence of a precisely defined interface control documents. In order to ensure that interfaces remain operational even if changes need to be introduced, a proper *configuration control* need to be applied for every interface. After all the elements have been properly assembled and tested in isolation the segment integration activity can start. The objective is to achieve that all the elements can communicate and interact correctly with each and are able to exchange all the signals as per the design documentation. Only at this stage can the ground segment be considered as fully integrated and ready to undergo the various verification and validation campaigns. A more detailed description of the assembly, integration, and verification (AIV) activities is provided in Chap. 2.

The ground segment systems engineer requires both the knowledge on how all the components have to interact with each other as well as a profound understanding of the functionalities and architecture of each single element. Chap. 4 intends to provide a first overview of the ground segment which is then followed by dedicated chapters for each of the most common elements.

The work of the operations segment is not so much centred on the development of hardware or software but rather based on processes, procedures, people, and skills. The main task is to keep the satellite alive and ensure that the active payload time to provide its service to the end user is maximised.[5] After the satellite has been ejected from the last stage of the launcher it needs to achieve its first contact with a TT&C station on ground as soon as possible. The establishment of this first contact at the predicted location and time is a proof that the launcher has delivered its payload to the foreseen location with the correct injection velocity. The important next steps are the achievement of a stable attitude, the deployment of the solar panels, and the activation of vital platform subsystems. This is very often followed by a series of orbit correction manoeuvres to deliver the satellite into its operational target orbit. All these activities are performed in the so called launch and early orbit phase (LEOP).

This phase is usually followed by the commissioning phase in which the payload is activated and undergoes an intense in-orbit test campaign. If this campaign is successful then the satellite can start its routine operations phase and provide its service to the end user. In each of these phases, operators must make intense use

---

[5] Especially for commercial satellites providing a paid service to the end user, any out of service time due to technical problems can have a significant financial impact for the company owning the satellite.

of the ground segment and the better it is designed and tailored to their specific needs, the more reliable satellite operations can be performed. It is therefore vital that the ground segment engineer considers the needs of the operators already in the early segment design phase as this will avoid the need for potential changes in later project phases, which always come at a considerably higher cost. A close cooperation between these two segments is therefore beneficial for both sides. The main operational phases, processes, and products are introduced in Chap. 15 and should help the ground segment engineer to gain a deeper understanding of this subject matter, allowing him to provide a more suitable and efficient segment design.

Every technical field comes with its own set of terms and acronyms which can pose a challenge to a freshman starting in a project. The aerospace field is notorious for its extensive use of acronyms which might get even worse once project specific terms are added on top. Care has been taken to only use the most common acronyms but to further facilitate the reading, a dedicated section with a summary of the most relevant terms and acronyms has been added.

Reference

1. International Council on Systems Engineering. (2011). *Systems engineering handbook*. A guide for systems life cycle processes and activities. INCOSE-TP-2003-002-03.2.2.

Chapter 2
Systems Engineering

The discipline of systems engineering (SE) can be understood as a work process that needs to be followed when defining, building, and delivering a complex system to an end user, properly considering all his needs and expectations. SE processes are to some extent applied in almost every engineering project and have therefore over the years been standardised by various international entities like the International Organisation for Standardisation (ISO), the International Electrotechnical Commission (IEC) [1], or the International Council of Systems Engineering (INCOSE) [2].[1] Whereas the ISO and INCOSE SE standards are not tailored to any technical domain, more specific descriptions for space projects have been derived by agencies like NASA [3] or ESA [4] to name only two of them.

The aim and justification for the application of SE principles is to manage engineering and project complexity in the most efficient way and to deliver the right product to the end user that fulfils all expectations. At the same time, SE reduces the risk that wrong decisions are being taken at an early project stage which could imply a significant cost impact later on in the project. This is demonstrated in Fig. 2.1 which depicts the actual life cycle cost (LCC) accrued over time and is based on a statistical analysis performed on a number of projects in the U.S. Department of Defense (DOD) [5]. As can be readily seen, even if only 20% of the overall LCC cost has been spend at the early development stage already more than 80% of the total LCC cost has been committed at this stage. A similar relationship exists for the cost impact to perform a design change which can either stem from a late introduction of new requirements or the need to correct a design flaw (see arrow in the centre of the figure).

The right amount of SE will clearly depend on the size, technical complexity, and contractual setup of the project. Whereas to little of SE leads to high variance in cost and schedule overrun, too much SE might put an unnecessary bureaucratic burden onto the project. Appropriate tailoring of the SE process is therefore required at the

---

[1] The INCOSE standard is based on the ISO/IEC/IEE15288:215 standard.

B. Nejad, *Introduction to Satellite Ground Segment Systems Engineering*, Space Technology Library 41, https://doi.org/10.1007/978-3-031-15900-8_2

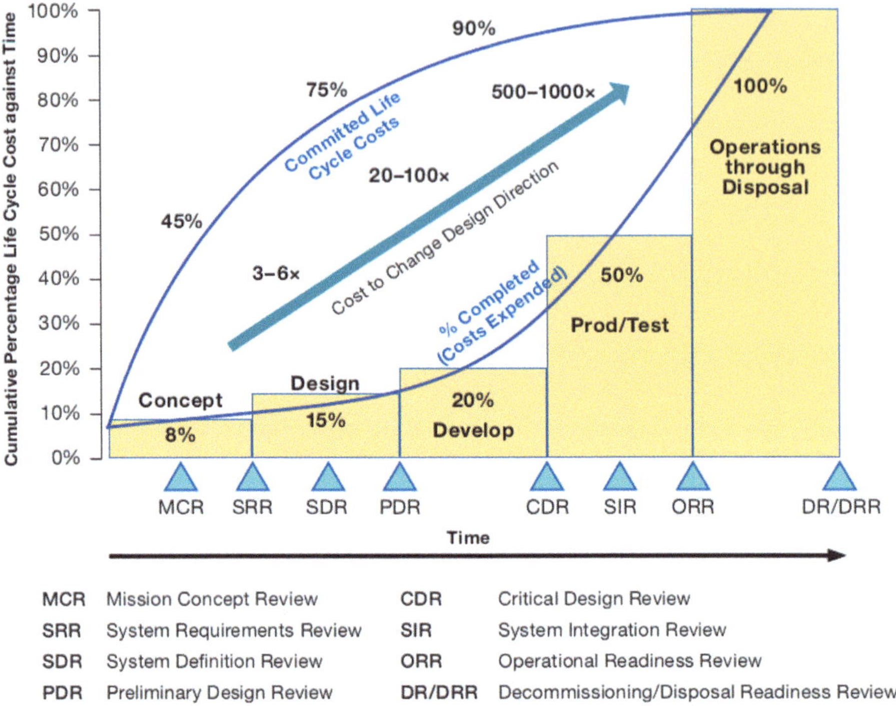

MCR	Mission Concept Review	**CDR**	Critical Design Review
SRR	System Requirements Review	**SIR**	System Integration Review
SDR	System Definition Review	**ORR**	Operational Readiness Review
PDR	Preliminary Design Review	**DR/DRR**	Decommissioning/Disposal Readiness Review

Fig. 2.1 Project life cycle cost versus time, reproduced Fig. 2.5-1 of SP-2016-6105 [3] with permission of NASA HQ

start of every project and should be clearly documented and justified in the *systems engineering management plan* (SEMP).

2.1   Project Planning

2.1.1   SOW, WBS, and SOC

Project planning by itself is quite a complex field and affiliated to the project management (PM) discipline. It is therefore typically run by specially trained personnel in a company or project referred to as project managers. Successful project planning is not an isolated activity but requires intense and regular cooperation with the SE domain, otherwise unrealistic assumptions in project schedules will be unavoidable and render them unachievable fairly soon. Especially at the beginning of a project a set of important planning activities need to be performed that are briefly outlined here as the project's system engineer will most likely be consulted.

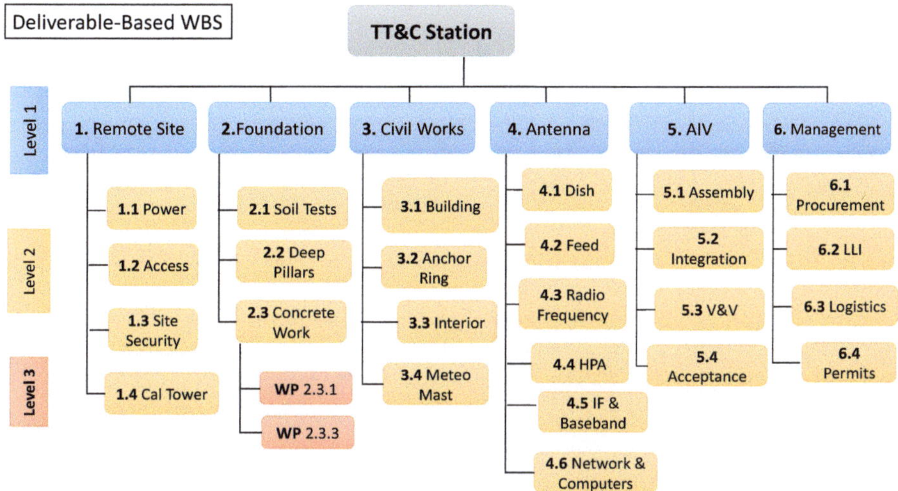

Fig. 2.2 Example of a deliverable-based work breakdown structure (WBS) for the case of the construction of a TT&C antenna. WP = Work Package, LLI = Long Lead Item, IF = Intermediate Frequency, Cal = Calibration

A project can only be achieved if the exact project scope is clearly understood, defined, verified, and controlled throughout the life cycle. This is defined as the *project scope management* activity by the Project Management Institute (PMI) (refer to PMBOK [6]). An important output of this activity is a statement of the project scope which is documented and referred to as the *statement of work* (SOW) and should describe in detail the project's expected deliverables and the work required to be performed to generate them. The SOW is a contractually binding document, therefore special attention must be given to reach a common understanding and document all the tasks in an unambiguous way with an adequate amount of detail that allows to verify whether the obligation has been fulfilled at the end of the contract. Having each task identified with an alphanumeric label (similar to a requirement) allows the contractor to explicitly express his level of contractual commitment for each task. The terms *compliant* (C), *partially compliant* (PC), or *non compliant* (NC) are frequently used in this context which the contractor states for each task in the SOW as part of a so called *statement of compliance* or SOC. Especially in case of a PC, additional information needs to be added that details the limitations and reasons for the partial compliance in order to better manage the customer's expectations.

Using the SOW definition of the project scope and all expected deliverables, the *work breakdown structure* can be generated which is a hierarchical decomposition of the overall scope into smaller and more manageable pieces [6]. There are two generic types of WBS in use, the *deliverable based* (or product-oriented) WBS and the *phase-based* WBS. An example of each type for the development and deployment of a TT&C station is shown in Figs. 2.2 and 2.3 respectively. For the

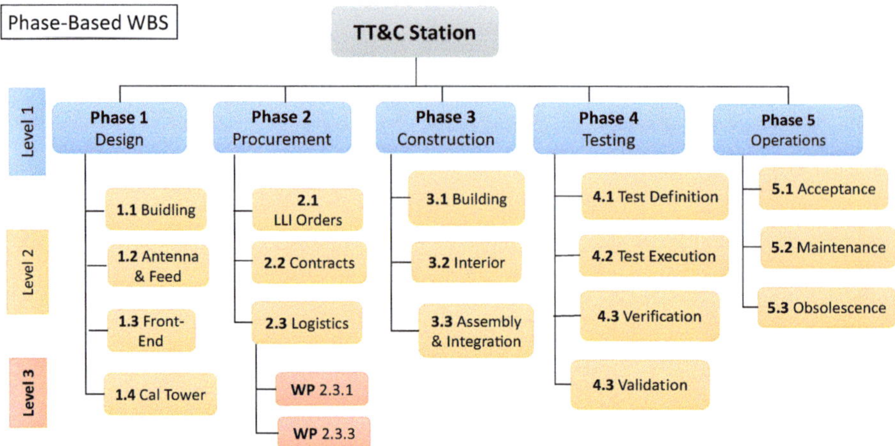

Fig. 2.3 Example of a phase-based Work Breakdown Structure (WBS) for the same case of the construction of a TT&C antenna

deliverable based WBS, the first level of decomposition shows a set of products or services, whereas in the phase based WBS the first layer defines a temporal sequencing of the activities during the project. In both cases each descending level of the WPS provides an increasing level of detail of the project's work. The lowest level (i.e., level 3 in the presented examples) is referred to as the *work package* (WP) level which is supposed to be the point at which the work can be defined as a set of work packages that can be reliably estimated in terms of complexity, required manpower, duration, and cost. A WP should always have a defined (latest) start time and an expected end time and identify any dependencies on external factors or inputs (e.g., the output of another WP providing relevant results or parts). The scope of work contained in a single WP should also be defined in a way it can be more easily affiliated to a dedicated WP responsible with a defined skill set. This can be either a company specialised on a specific product or subject matter, a specialised team in a matrix organisation, or even a single person.

Once all the WPs have been identified as part of a WBS and contracts have been placed to WP responsible, the *organisational breakdown structure* (OBS) can be generated. This represents a hierarchically organised depiction of the overall project organisation as shown in Fig. 2.4 showing an OBS example for a typical space project with three contractual layers. The OBS is an important piece of information for the monitoring of the overall project progress which requires the regular reception of progress reports, the organisation of progress meetings, and the assessment of achieved milestones. The OBS is also an important input for the financial project control as it clearly identifies the number of entities to which a formal contractual relationship needs to be established and maintained throughout the project. This requires the regular initiation of payments in case milestones have been achieved or other measures in case of partial achievements or delays.

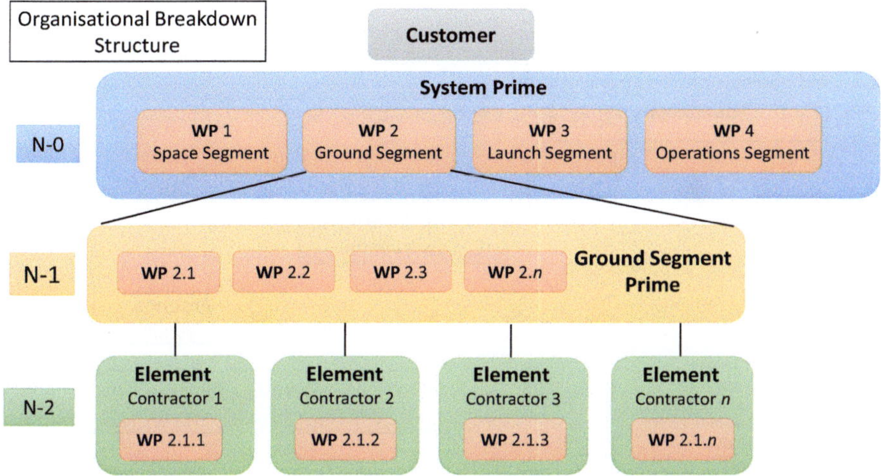

Fig. 2.4 Example of an organisational breakdown structure (OBS) showing three contractual levels below the customer

2.1.2 Schedule and Critical Path

One of the most fundamental aspects of the project is its schedule that defines the start and predicted end time. An initial schedule is usually expected prior to the kick-off of a project and will serve as an important input to receive its approval and funding from the responsible stake holders. It is obvious that such an early schedule must be based on the best estimate of the expected duration of all the WPs, considering all internal and external dependencies, constraints, and realistic margins. Throughout the project execution, a number of schedule adjustments might be needed and need to be reflected in regular updates of the *working schedule*. These needs to be carefully monitored by both the project responsible to take appropriate actions and to inform the customer. It is important to keep in mind that both schedule and budget are always interconnected, as problems in either of these domains will always impact the other one. A schedule delay will always imply a cost overrun, whereas budget problems (e.g., cuts or delays due to approval processes) will delay the schedule.[2]

In order to generate a project schedule the overall activities defined in the SOW and WBS needs to be put into the correct time sequence. This is done using the *precedence diagramming method* (PDM) which is a graphical presentation that puts all activities into the right sequence in which they have to be executed. For a more accurate and complete definition as well as tools and methods to generate

---

[2] Especially in larger sized projects the latency in cash flow between contractual layers can introduce additional delays and could even cause existential problems for small size companies that have limited liquidity.

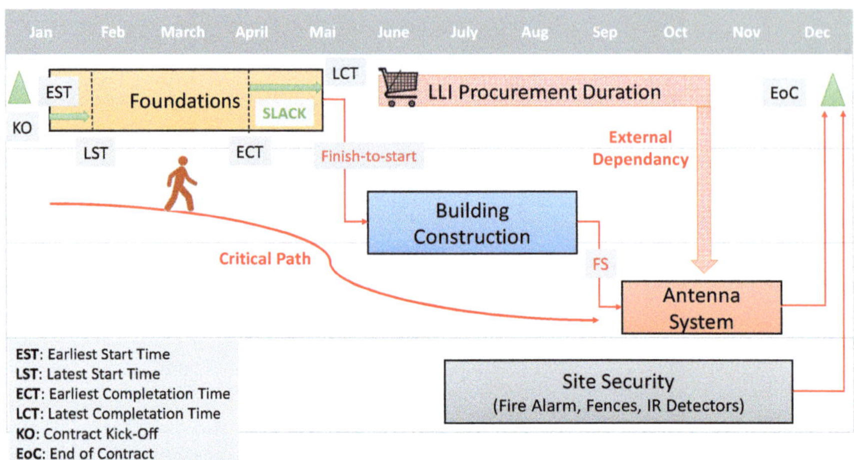

Fig. 2.5 Simplified version of a precedence diagram for a TT&C antenna

such diagrams the reader should consult more specialised literature on project management (e.g., [6] or [7]). An example of a precedence diagram is depicted in Fig. 2.5 which shows in a very simplified manner the construction of a TT&C antenna with an overall project duration of only one year and four main tasks to be executed between kick-off (KO) and end of contract (EoC). To allow proper scheduling, each task needs to be characterised by an earliest start time (EST), latest start time (LST), earliest completion time (ECT) and latest completion time (LCT) which are marked on the first task in Fig. 2.5. The most prevalent relationship between tasks in a precedence diagram is the *finish-to-start* relationship which demands that the successor activity can only start if its designated predecessor one has been completed. The difference between the earliest and latest start or completion times is called *slack* and defines the amount of delay that an activity can build up (relative to its earliest defined time) without causing the depending successor to be late.[3]

In the presented example a finish-to-start dependency is indicated via a red arrow that points from the end of one task to start of its dependent successor. This applies for the preparation of the foundation which obviously needs to be completed before the construction of the antenna building can start. The same is the case for the building construction which requires to be finalised in order to accommodate the antenna dish on top and allow the start of the interior furbishing with the various antenna components (e.g., feed, wave guides, BBM, HPA, RF components, ACU, computers, etc.).

---

[3] Schedule slack is an important feature that can improve schedule robustness by absorbing delays caused by unpredictable factors like severe weather conditions, logistics problems (e.g., strikes or delay in high sea shipping), or problems with LLI procurement.

It is important to understand that such finish-to-start dependencies are highly important players in project planning as they identify the work packages and activities that need to be considered as schedule *critical*. This means that any delay of such a critical activity has a potential impact on the overall completion date of the project. The alignment of all critical activities is referred to as the *critical path* and requires the highest attention when monitoring project progress. If an activity on the critical path has problems to finish in the foreseen time (i.e., its designated LCT cannot be met anymore), a project schedule alert needs to be raised and immediate action taken to minimise that delay. The site security task in Fig. 2.5 does not have a finish-to-start dependency to any of the other activities and can therefore be considered to be *off* the critical path.[4]

Another important activity that needs to be initiated well in advance, is the procurement of so called *long lead items* (LLIs). This comprises any equipment or material that requires an extensive amount of time from the placement of order to actual delivery. Even if such LLI orders are placed in time considering the average procurement lead time from previous experience, unexpected delays can always occur (e.g., a sudden market shortage due to high demand or a crisis, logistics problems due to strikes, etc.). This is the reason why LLIs are usually categorised as an external dependency that requires close monitoring and should even be tracked as a project risk.

2.1.3   Project Risk

Project management needs to deal with the identification of project risk, its categorisation, and monitoring throughout the entire life cycle. Project risks (also referred to as threats) can be seen as any event that can imply a negative or undesirable[5] impact on the overall project execution and in a worst case scenario even cause project failure. A first analysis of potential risks should be performed at the very start of a project and is usually documented in the so called *risk register*. Such a register usually categorises the identified risks into four types, technical, cost, schedule, and programmatic (cf. [2]). Technical risks can affect the ability of the system to meet its requirements (e.g., functional, performance, operational, environmental, etc.). Cost and schedule risks imply overrun in budget or delay in the achievement of project milestones. Programmatic risks are external events that usually lie beyond the control of the project manager as they depend on decisions

---

[4] The reason is that the implementation of the site security related installations (e.g., fire detection and alarm, fences) can be performed in parallel to the other activities and need to be finalised only at the end of the project.

[5] If the impact of an event is desirable, it is termed *opportunity*.

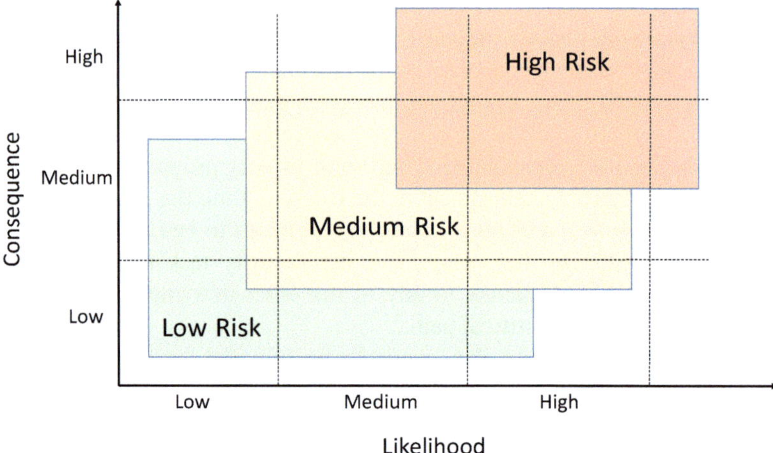

Fig. 2.6 Risk severity measurement assessment based on the product of the likelihood of occurrence and the projected consequence this event could happen

taken at a higher level. Examples are legal authorisation processes or funding cuts due to a changing political landscape.[6]

Once all possible risk events have been identified and categorised, they need to be measured as high, medium, or low project risk. The risk level is usually measured based on the product of two factors: (1) the likelihood or probability that event occurs, and (2) the consequence or damage it can cause. This is usually presented in a two-dimensional diagrams as shown in Fig. 2.6. Both the number of identified risks as well as their respective severity level can change during the project life cycle which requires a regular update and delivery of the risk register at major project milestones. The ultimate goal of risk management is to move risks from the high into the medium or even low risk area of Fig. 2.6. This can only be achieved if the source and background of each risk item is well understood as this allows to put the right measures into place at the right time.

2.2 System Hierarchy

The creation of system hierarchies is a very common process in engineering and refers to the partitioning of a complex system into smaller more manageable entities or subsystems. The formation of system hierarchies not only benefits the organisation of the work at technical level but also the overall project management

---

[6] Larger scale NASA space projects with development times spanning several years can be adversely affected by the annual (fiscal year) budget approval by the United States Congress.

process as it might support the definition of the WBS and OBS as discussed earlier in the text.

In a system hierarchy, each subsystems is supposed to perform a very specific set of functions and fulfil a set of requirements that has been made applicable to it (or *traced* to it). Any subsystem can be further decomposed to the components or parts level if needed. Basic examples of such hierarchies are given for the Space Segment and the Ground Control Segment in Chaps. 3 and 4 respectively. To find the right balance in terms of decomposition size and level, one can use the 7 ± 2 elements rule which is frequently worded in SE literature as a guideline (cf. [1]).[7] Once an adequate decomposition has been found and agreed it needs to be formally documented in the so called *product tree* which serves as an important input for the development of the *technical requirements specification* (TS). In this process top level requirements are decomposed into more specific and detailed ones that address lower level parts of the system. The outcome of this is reflected in the *specification tree* that defines the hierarchical relationship and ensures forwards and backwards traceability between the various specification levels and the parts of the system they apply to (refer to [8]).

2.3 Life-Cycle Stages

The overall management of a technical project is usually based on the definition and execution of specific life cycle stages which allows to distinguish the various development phases that a system has to go through in order to mature from the user requirements definition to the final product. A life cycle model is a framework that helps to define and document the complete set of activities that need to be performed during each stage and defines the exact conditions that have to be met before the subsequent stage can start. In addition, characteristic interactions and iterative mechanisms can be more easily defined and exercised. The decision to proceed from one stage to the next is based on the assessment of the proper completion of all the required tasks for that stage. This is usually performed in the context of a so called *decision gate* which is also referred to as a *project milestone*, *review*, or *key decision point*.

Figure 2.7 provides an overview of the life cycle stages as defined by three different organisations, i.e., INCOSE (upper panel), NASA (middle panel), and ESA (lower panel). One can readily recognise similar approaches and terminology used for the life cycle stages and the timing of important review milestones. The early stages are mainly focusing on the identification, clarification, and wording of stakeholder requirements. This is a very important activity as its outcome will have a significant impact on the system design which is supposed to be provided

---

[7] This heuristic rule suggests that at each level any component in a system should not have more than 7 ± 2 subordinates.

INCOSE (ISO/IEC/IEEE 15288:2015)				Utilization	Retire-
Concept	Development	Production		Support	ment

NASA Formulation		Approval		Implementation			
Pre-Phase A	Phase A	Phase B	Phase C	Phase D	Phase E	Phase F	
Concept Study	Concept & Technology Study	Preliminary Design &Technology Completion	Final Design & Fabrication	System Assembly, Integration & Test, Launch	Operations & Sustainment	Closeout	
MCR SRR SDR		PDR	CDR SIR	SAR ORR LRR FRR		DR DRR	

ESA (ECSS-M-ST-10C)						
Phase 0	Phase A	Phase B	Phase C	Phase D	Phase E	Phase F
Mission analsyis & needs identification	Feasibility	Preliminary definition	Detailed Definition	Qualification & Production	Operations & Utilization	Disposal
MDR PRR		SRR PDR	CDR	QR	AR FRR CRR ORR LRR ELR	MCR

Fig. 2.7 Comparison of life cycle stage definition and review milestones of three different entities: international council of systems engineering [2] (upper panel), NASA [3] (middle panel), and the European space agency [9]. The 3 letter acronyms refer to review milestones (refer to Acronyms list)

in the subsequent stages, i.e., a preliminary design in phase B and the detailed design in phase C. Once the design is agreed and *baselined*,[8] the production and qualification can start (phase D). The qualification process itself comprises a verification, validation, and acceptance activity. The operations phase (Phase E) in space projects usually starts with the launch of the space vehicle and comprises a LEOP and in-orbit commissioning phase prior to the start of routine operations. The final stage starts with the formal declaration of the end-of-life of the product and deals with the disposal of the product. In a space project this usually refers to the manoeuvring of the satellite into a disposal orbit or its de-orbiting.

Figure 2.7 also shows the various review milestones (refer to 3-letter acronyms) that are scheduled throughout the life cycle. These reviews can be considered as decision gates at which the completion of all required activities of the current phase are being thoroughly assessed. This is followed by a board decision on whether the project is ready to proceed to the subsequent stage, or some rework in the current one is still needed. As such reviews might involve external entities and experts, it is very useful to provide a review organisation note that outlines the overall review objectives, the list of documents subject to the review, the review schedule, and any other important aspects that need to be considered review process (e.g., the

---

[8] This implies that any further change needs to follow the applicable configuration control process.

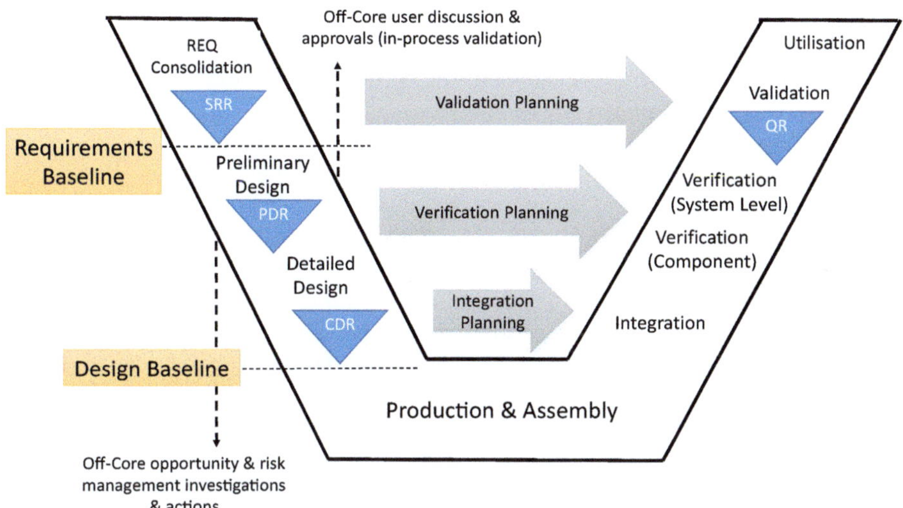

Fig. 2.8 The Vee Systems Engineering model with typical project milestones as defined in [11]

method how to raise a concern or potential error as a *review item discrepancy* (RID), important deadlines, etc.).[9]

Figure 2.8 shows the life-cycle stages in the so called Vee model [11, 12] named after the V-form shape representation of the various SE activities. Starting the SE process from the top of the left branch, one moves through the concept and development stages in which the requirements baseline must be established and a consistent design derived from it. Entering the bottom (horizontal) bar, the production, assembly, and integration activities take place, which deliver the system of interest at the end. Entering the right side of the "V" (from the bottom moving upwards) the various qualification stages are performed, starting with the verification ("Did we build it right ?"), the validation ("Did we build the right thing ?"), and finally the product acceptance ("Did any workman-ship errors occur ?"). The idea to represent these in a "V" shape is to emphasise the need for interactions between the left and the right sides. The elicitation and wording of the high level requirements (top left of the "V") should always be done with the utilisation and validation of the system in mind. The utilisation or operational concept of a system will impact both the high level user requirements and the validation test cases. The specification stages that aim to define the requirements need to also define the verification methods and validation plans. This will avoid the generation of either

---

[9] Guidance for the organisation and conduct of reviews can be found in existing standards (see e.g., ECSS-M-ST-10-01C [10]).

inaccurate or "impossible to verify" type of requirements that can imply design flaws in the system.

Another important aspect is shown by the vertical lines on the left side of the "V" which represent the so called *of core* activities, as they are initiated in parallel to the sequential (in core) ones. The downward directed arrow shows the activities related to opportunity and risk management activities where opportunity in this context refers to the consideration of alternative design solutions in order to improve system robustness or take into account new technologies. Risk in this context refers to the critical analysis of potential heritage design decisions or components, in case these have been adopted for the new design.[10] The upward directed arrow shows the off-core activities that refer to the continuous involvement of the customer and any relevant stakeholder in the proposed design baseline definition which is an important activity as it will reduce the likelihood for change requests in late stages of the project.

2.4   Life-Cycle Models

2.4.1   *Sequential or Waterfall*

The classical approach for SE life-cycle models are *plan-driven* or *sequential* models which due to their strictly one directional nature are also referred to *waterfall* models and have the following typical characteristics (cf. [12]):

- A strict adherence to a pre-defined sequence of life-cycle stages and processes with a detailed set of tasks expected to be performed within each of the defined stages.
- A strong focus on the completeness of documentation and its proper configuration control.
- An explicit need for a detailed requirement specification from the very beginning that has a clear traceability to the system design and qualification documents.
- The application of established standards throughout all the project phases.
- The use of measurement parameters and success indicators for project control (e.g., KPIs, milestone payment plans, etc.)

Due to their predictability, stability, and repeatability the plan-driven or waterfall models are more frequently applied in larger scale projects due to their more complex contractual set ups involving a multitude of subcontractors. Sequential

---

[10] An example for this is the failure of the Ariane 5 maiden flight on 4 June 1996 which could be traced to the malfunctioning of the launcher's inertial reference system (SRI) around 36 s after lift-off. The reason for the SRI software exception was an incorrect porting of heritage avionic software from its predecessor Ariane 4 which was designed for flight conditions that were not fully representative for Ariane 5 (see ESA Inquiry Board Report [13]).

methods are also preferred for the development of safety critical products that need to meet certification standards before going to market (e.g., medical equipment).

2.4.2 Incremental Models: Agile, Lean and SAFe®

Incremental and Iterative Development (IID) methods are suitable when an initial system capability is to be made accessible to the end user which is then followed by successive deliveries to reach the desired level of maturity. Each of these increments[11] by itself will then follow its own mini-Vee development cycle, which can be tailored to the size of the delivery. This method has the advantage to not only accelerate the "time to market" but also allows to consider end user feedback when "rolling out" new increments and change the direction in order to better meet the stakeholder's needs or insert latest technology developments. This could also be seen as optimising the product's time to market in terms of the vectorial quantity *velocity* rather than its scalar counterpart of *speed*, with the advantage that the vectorial quantity also defines a direction that can be adapted in the course of reaching its goal. Care must however be taken to properly manage the level of change introduced during subsequent increments in order to avoid the risk of an unstable design that is continuously adapted in order to account for new requirements that could contradict or even be incompatible with previous ones.

IID methods have already been extensively used in pure software development areas, where *agile* methods are applied since much longer time. This is mainly due to the fact that pure software based systems can be more easily modified, corrected, or even re-engineered compared to systems involving hardware components. The latter ones usually require longer production times and might run into technical problems if physical dimensions or interfaces change too drastically between consecutive iterations. More recently agile development methods have also gained more significance in the SE domain, mainly through the implementation of the following practices (cf., [14]):

- Perform short term but dynamic planning rather than long term planning. Detailed short term plans tend to be more reliable as they can be based on a more realistic set of information whereas long term plans usually get unrealistic soon, especially if they are based on inaccurate or even wrong assumptions taken at the beginning of the planning process. This approach of course demands the readiness of project managers to refine the planning on a regular basis throughout the project, typically whenever more information on schedule critical matters becomes available.
- Create a verifiable specification that can be tested right at time when being formulated. In order to verify a requirement or specification it has to first be

---

[11] In the agile Scrum framework, increments are also referred to as "sprints".

transformed into an executable (testable) model. This is clearly different from the traditional SE approach where requirements are usually captured in a pure textual set of statements, accepting the fundamental weakness of language for its lack of precision and ambiguous interpretation. This new approach to use verifiable specification is a common method applied in *model based systems engineering* (MBSE) which will be outlined in more detail in the subsequent section. The major advantage of verifying a requirement right after its generation is to immediately check whether it is precise, consistent, unambiguous, and fulfilling its objective. If this is not the case it can be corrected and improved *before* a system design is derived from it. This will favour that the final derived system design meets the customer's needs and avoids the need for expensive modifications in late project phases.

- Apply *incremental* approaches as much as possible which allow to verify the quality of a work product at an early stage through the consideration of user feedback that can serve to either correct or improve the subsequent increment.
- Implement a continuous verification approach at all levels starting with the requirements specification at the project start (see bullet above), followed by unit verification for all relevant software and hardware components, integration level after the components have been connected via the foreseen interfaces, and finally at system level once the entire product is assembled. In practice this means to follow a *test driven development* (TDD) approach, in which test scenarios (and procedures) for a product are generated in parallel to the product itself. This allows testing to be applied as soon as a unit or component has been generated favouring the early discovery and correction of errors.

A more recent product development and delivery philosophy is referred to as the *Scaled Agile Framework* (SAFe®) [15] and aims to combine the agile (incremental) production techniques outlined above with business practices put forward by the *Lean* principles [16]. The lean thinking is rooted in the Toyota "just-in-time" philosophy [17] and identifies activities or products that can be considered as value (i.e., adding benefit to the product) or waste (i.e., having a negative impact like increasing cost or delay the schedule). The objective is to optimise work processes by increasing value and reducing (or even removing) waste. The combined view is referred to as the *Lean-Agile* mindset that represent the core values and principles of SAFe®.

Increments are delivered by so called multidisciplinary "Agile Teams"[18]. As usually a broad range of speciality skills is needed that cannot be contained within a single Agile team, multiple of such (cross functional) teams must collaborate which then form a so called *Agile Release Train* (ART) [19] which engages in continuous exploration, integration, deployment, and customer release activities. The ART can be seen as an alternative description of an incremental delivery approach, where the train synergy is used to emphasise a fixed start time (train departure) and a fixed delivery velocity defined by the overall project schedule. The train cargo would relate to the new functionality or features that is supposed to be provided by that given increment. The SAFe® framework furthermore integrates and combines concepts from *Scrum, Extreme Programming*, and *Kanban* as key technologies.

Scrum is a "lightweight framework" that originates from the software development world and was defined by Jeff Sutherland and Ken Schwaber [20]. A Scrum development builds on small teams (typically ten or even fewer members) that comprise the *product owner* (i.e., the product's stakeholder), the *scrum master* (i.e., the leader and organiser of the scrum team), and the *developers*. The incremental delivery is referred to as a *sprint* and is characterised by a predefined sprint goal (selected from a *product backlog*[12]) and a fixed duration of maximum one month or even less. The outcome of a sprint is inspected in a so called *sprint review*.

Extreme programming is an agile methodology originally formulated by Kent Beck as a set of rules and good practices for software development which have subsequently been published in open literature [21]. The extreme programming rules address the areas of both coding and testing and introduce practices like pair programming,[13] test-driven development, continuous integration, strict unit testing rules,[14] simplicity in design, and sustainable pace to control the programmer's work load.

The Kanban method is a scheduling system that has its origins in the car industry (e.g., Toyota) and defines a production system that is controlled and steered by requests that actively steer the stock or inventory resupply and the production of new items. Such request can be communicated by physical means (e.g., Kanban cards) or an electric signal and are usually initiated by the consumption of a product, part, or service. This allows an accurate alignment of the inventory resupply with the actual consumption or demand of an item. SAFe® suggests Team Kanban as a working method for the so called *System Teams* [23], which are responsible for the build up of the agile development environment as well the system and solution demos. The latter one is a demonstration of the new features of the most recent increment to all relevant stakeholders, allowing to receive immediate feedback. Important working methods in the Kanban context are a continuous monitoring and visualisation of workflows as well as the establishment of *work in process* (WIP) limits and their adaption if necessary. A graphical technique to visualise workflows is shown in the Kanban board depicted in Fig. 2.9. Work items shown as green squares are *pulled* from the *team backlog* on the left from where they move through the various workflow states, "Analyse", "Review", "Build", and "Integrate/Test". The Analyse and Build states have buffers, which allows to "park" a work item when it has been completed until it is ready to enter the next one. This is beneficial in case the next

---

[12] The product backlog is defined in [20] as an emergent, ordered list of what is needed to improve the product and is the single source of work undertaken by the scrum team.

[13] Pair programming (also known as collaborative programming) is a coding methodology that involves two developers collaborating side-by-side as a single individual on the design, coding and testing of a piece of software: one controls the keyboard/mouse and the other is an observer aiming to identify tactical defects and providing strategic planning [22].

[14] E.g., all code must undergo successful unit testing prior to release.

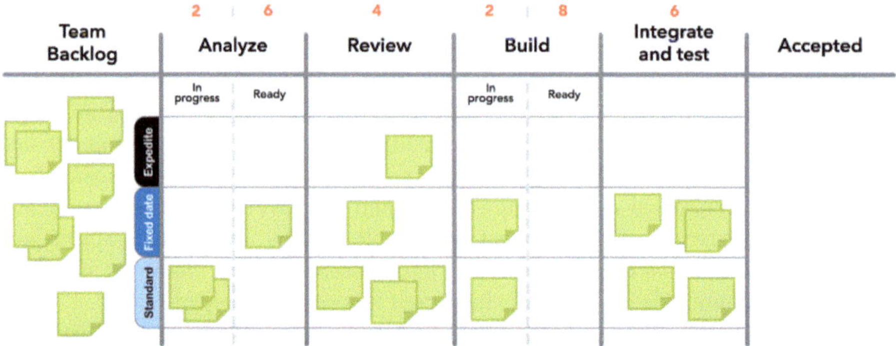

Fig. 2.9 Kanban board example, reproduced from [24] with permission of Scaled Agile Inc

step in the process depends on external factors that are not directly under the control of the project like for example external reviewers or the availability of a required test infrastructure. If maintained on a regular basis, the Kanban board supports a continuous progress monitoring, which is referred to as *measuring the flow* in SAFe® terminology.

2.5   Model Based Systems Engineering (MBSE)

In a traditional SE framework all of the engineering data like requirements specification, architectural design descriptions, interface control documents, verification and validation plans, procedures, trade studies, etc. are distributed over a multitude of documents and captured as textual specifications or possibly as schematic drawings.[15] As any description in textual language will contain a certain level of imprecision and allow multiple interpretation (often simply depending on the reader and his or her background), requirements and architectural design derived from it can contain significant errors. These are however usually only discovered at a later stage in the project life cycle when corrections or modifications usually imply a considerable higher amount of work which usually implies delays in the project schedule and additional cost.

In recent years it has therefore been realised that the use of models allows to construct an executable specification and system architecture which can be verified immediately after its creation which led to the term *model based systems engineering* or MBSE. The language used to formulate these models is called the *Systems Modeling Language* or briefly SysML [25] and integrates many building blocks of the well known Unified Modeling Language (UML) also developed by

---

[15] The traditional SE approach is therefore also sometimes referred to as *textual based SE approach.*

the Object Management Group [26]. The use of models allows to create, manage, and verify engineering data that is less ambiguous, more precise, and most important consistent. In contrast to having a multitude of documents like in the textual-based SE, the engineering data is stored in a single model repository which is used to construct detailed views of the contained information in order to address specific engineering needs or to perform a detailed analysis. Even if different views need to be generated in order to provide the required inputs to the various engineering domains, their consistency can always be easily guaranteed as they were derived from the same source of data. This methodology also facilitates a fast and more accurate impact analysis of a technical decision or design modification on the overall system and its performance, which otherwise would be considerably more involved to generate and potentially even less accurate.

The SysML language provides a set of diagrams that allow to capture model characteristics, its requirements, but most important specific behavioural aspects like activities (i.e., transition of inputs to outputs), interactions in terms of time-ordered exchange of information (sequence diagrams), state transitions, or constraints on certain system property values (e.g., performance or mass budget). The advantage to use a well defined (and standardised) modelling language is the compatibility between various commercially available tools allowing to more easily exchange modelling data between projects or companies.

2.6   Quality Assurance

The last section of this chapter describes necessary administrative activities and practices which are grouped under the terms *quality assurance* (QA) or *product assurance* (PA) and aim to ensure that the end product of a developing entity fulfils the stakeholder's requirements with the specified performance and the required quality. Quality in this context refers to the avoidance of workmanship defects or the use of poor quality raw material. In order to achieve this, the developing entity usually needs to agree to a set of procedures, policies, and standards at the project start, which should be described in a dedicated document like the *Quality Assurance Management Plan* (see e.g., [27]) so it can be easily accessed and be followed during the various project phases.

2.6.1   Software Standards

In today's world software is almost present in any technical device or component and its development has therefore gained significant importance over the last decades. Even areas that were previously entirely dominated by pure hardware solutions are today steered to some extent by software programmes. One reason is the triumphant advance of field-programmable gate arrays (FPGAs) which are

Table 2.1 Software design assurance level (DAL) as specified by DO-178B [28]

Failure condition	Software level	Description	Failure probability (P)
Catastrophic	A	A failure may cause multiple fatalities, usually with loss of the airplane	$P < 10^{-9}$
Hazardous/ severe-major	B	Failure has a large negative impact on the safety or performance, reduces the ability of the crew to operate the aircraft, or has an adverse impact upon the occupants	$10^{-9} < P < 10^{-7}$
Major	C	Failure slightly reduces the safety margin or increases the crew's workload and causes an inconvenience for the occupants	$10^{-7} < P < 10^{-5}$
Minor	D	Failure slightly reduces the safety margin or increases the crew's workload and causes an inconvenience for the occupants	$10^{-5} < P$
No effect	E	Failure has no effect on the safe operation of the aircraft	N/A

integrated circuits that can be programmed using a hardware description language. This has lead to innovations like software-defined radios, networks, or even satellite payloads.

As long as software is developed by human beings it will to some extent contain software flaws that are introduced during its design or generation phase. If such flaws (colloquially referred to as "bugs") are not discovered and corrected prior to the software release they will affect the functionality or performance of the final product. Additional software patches will then be required to corrected them. The application of *software standards* during the development, testing, and deployment phase are an important means to minimise such software bugs and the need for later corrections. Achieving the right balance between the application of a software standard and the need for agility during the development phase can sometimes be quite challenging, especially for a field where fast time to market is more demanding than for any other engineering discipline.

A proper tailoring of the applied standard to the project needs is therefore a key aspect to achieve this. One important consideration is the evaluation of the safety criticality of a software component which is a measure of the overall impact in case of its failure. This is of special importance for software components of airborne systems that deliver human passengers. Aircraft certification agencies like the U.S. Federal Aviation Administration (FAA) or the European Organization for Civil Aviation Equipment (EUROCAE) have therefore adapted the *Design Assurance Level* (DAL) categorisation shown in Table 2.1 which is defined in the standard DO-178B/EUROCAE ED-12B [28] and used for the certification of all software used in airborne avionics systems.[16] Depending on the DAL level a specific software component falls into, a different set of assurance requirements (objectives) has to be fulfilled during the design, coding, and verification stages and a different failure probability must be demonstrated which is indicated in the most right column of Table 2.1.

Especially for space projects that provide end user services applied for life critical applications or the civil aviation industry itself (e.g., navigation services for automated aircraft landing), the adoption of the DAL concept and its requirements in the project specific software standard is highly recommended as this will simplify required certification processes. Software standards should provide clear guidelines for at least the following areas:

- A clear description of the software development life cycle to be applied (e.g, waterfall, incremental, etc) and the interaction with a corresponding model at a hierarchically higher level (e.g., software versus element level, or element versus segment level). Software life cycles frequently use a very similar terminology for the definition, development, and qualification phases as corresponding review milestones at the next higher level. The use of prefixes like "SW-PDR", "SW-DDR", or "SW-CDR" can therefore be useful to easier distinguish them. It is also important to provide guidelines on the correct (and meaningful) synchronisation or combination of certain review milestones for the sake of efficiency. As an example, a SW-CDR could be combined with its higher level Element-CDR in case both software and hardware have been developed in parallel and could reach the same level of maturity at the same point in time.
- The definition and description of coding standards, rules, and practices that have to be applied for any new software component being developed. This could comprise the specification of allowed programming or scripting languages, the definition of header layouts and required information contents for source code files, the specification of programming rules and policies (e.g., dynamic memory allocation and de-allocation), and recommendations for coding styles. Another aspect that should be addressed in this context is the definition of rules on how an application software has to deal with errors. This should comprises both the logging of errors (and maintenance of log files) and the prevention of fault propagation.
- Software qualification requirements define the type and level of testing that needs to be performed in order to fully qualify any new developed application. The smallest software module that can be identified by a compiler in an application is referred as a *Software Unit* (SU) which is schematically depicted in Fig. 2.10 where two application programs are run on the same platform operating

---

[16] The DO-178B standard was actually developed by the Radio Technical Commission for Aeronautics (RTCA) in the 1980s and has more recently been revised (refer to DO-178C [29]) in order to consider modern development and verification techniques used in model-based, formal, and object-oriented development [30].

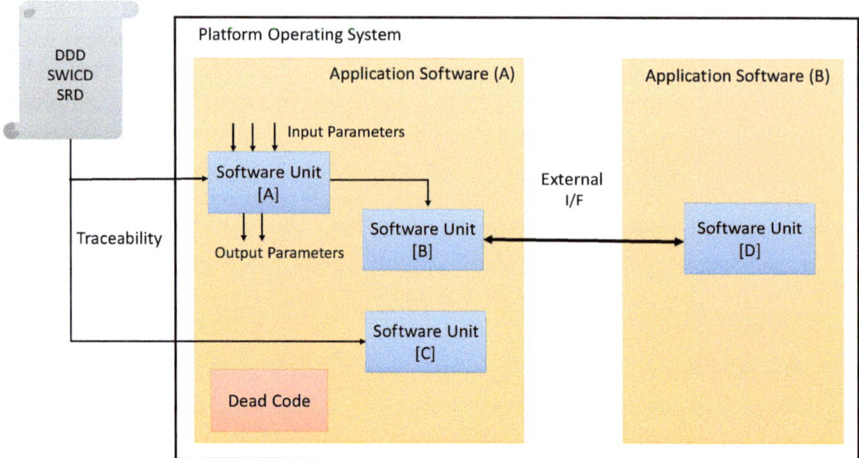

Fig. 2.10 Software testing is performed at unit, application, and interface level. Traceability between code and the applicable documentation needs to be demonstrated in order to avoid dead code. DDD = Detailed Design Document, SWICD = Software Interface Control Document, SRD = Software Requirements Document

system.[17] Unit testing is the lowest level of software testing and is supposed to demonstrate that the SU performs correct under all given circumstances. This is usually achieved via automated parameterised testing in which the SU is called multiple times with different sets of input parameters in order to demonstrate its correct behaviour under a broad range of calling conditions. Higher level testing needs to be performed at application level and needs to demonstrate the its proper functionality which must be traceable to the relevant design document and the applicable software requirement (SRD). Executable code in reused software components that is never executed and cannot be traced to any requirement or design description has to be identified as "dead code" and is subject to removal unless its existence can be justified.[18] Another important testing activity is interface testing which has to demonstrate the correct exchange of all interface signals (in terms of format, size, file name, protocol, latency etc.) as per the applicable ICDs.

---

[17] This can either be the OS of the host computer or a the one from a Virtual Machine (VM).

[18] The detection of dead code can be achieved by either source code inspection or is the outcome of structural coverage measurement campaign.

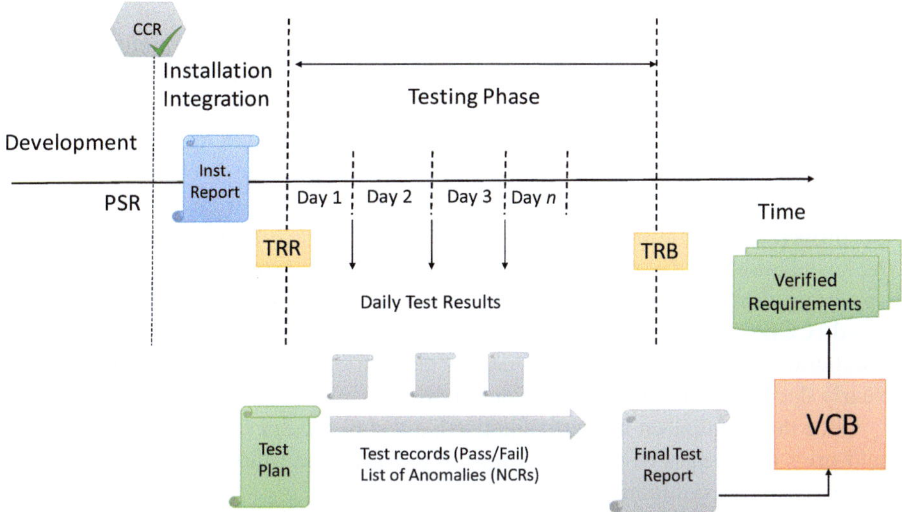

Fig. 2.11 Test Campaign Planning: PSR = Pre-shipment Review, TRR = Test Readiness Review, TRB = Test Review Board, CCR = Configuration Change Request, VCB = Verification Control Board

2.6.2   Test Campaign Planning

Quality Assurance standards also play an important role for the proper planning and execution of testing activities during the verification and validation phases of a project. The correct execution and documentation of a test campaign can maximise its efficiency and outcome and avoid the need for potential test repetitions at a later stage. The schematic in Fig. 2.11 depicts an example of a test campaign planning scheme in which the various stages are depicted on a horizontal timeline. A prerequisite of any testing activity is the finalisation of the development of a certain part, software component, or sub-system which is ready to be handed over to a higher level recipient. An example could be something simple like the ground segment receiving a software maintenance release or a patch or something more complex like a remote site receiving a brand new antenna dish to be mounted on the antenna building structure. Any such transfer needs to fulfil a certain set of prerequisites which need to be verified in a dedicated review referred to as the *pre-shipment review* (see e.g., [31]). Such a PSR is held to verify the necessary conditions for the delivery ("shipment")[19] from the developing to the recipient entity. The conditions to authorise this shipment usually comprise the

---

[19] The term shipping her refers to any kind of transfer, be it an electronic file transfer, postal transfer of media, transfer via road or railway, or even a container ship crossing the ocean.

readiness of the destination site to receive the item,[20] the readiness of the foreseen transportation arrangements and transportation conditions (e.g., mounting of shock sensors for sensitive hardware, security related aspects that impact the delivery method, the exact address and contact person at destination site, etc.), the availability of installation procedures, and the completeness of installation media.

After the successful shipment the installation and integration of the new component can be performed. In case this implies a modification of an existing baseline, the proper configuration control process needs to be followed. In other words, the installation of the element is only allowed if a corresponding *configuration change request* or CCR has been approved by the responsible configuration control board (CCB). The installation itself needs to follow the exact steps defined in the installation procedures and any deviations or additional changes required need to be clearly documented in the installation report.

Once the installation has been finalised, a *test readiness review* (TRR) should be held prior to the start of test campaign. In the TRR the following points should be addressed and documented as a minimum:

- Review of the installation report to discuss any major deviations and their potential impact.
- Define the exact scope of the test campaign, outlining the set of requirements or the operational scenario that should be validated.
- The readiness of all required *test procedures* (TPs) should be confirmed and the procedures should be put under configuration control. The TPs should also contain the traces of the requirements to the test steps where the verification event occurs. If needed, a test procedure can be corrected or improved during the test run, but any performed change should be clearly identified for incorporation into future versions of the TP (and re-used in future campaigns if needed).
- The test plan should identify the detailed planning of the campaign. The plan should detail the number of tests that are scheduled for each day and identify the required resources. Resources should be considered from a human point of view (test operators and witnesses) and machine point of view (test facilities and required configuration).
- In order to be reproducible, the exact configuration baseline of the items under test as well as the used test facilities need to be identified and documented. Either the configuration is specified in detail in the *minutes of meetings* that are generated during this review or a reference to another document is provided.[21]

During the testing activity, the detailed test results need to be properly documented. Each performed test step needs to have a clear indication whether the system has performed as expected (pass) or any problems have been observed (fail).

---

[20] For dust sensitive components like a satellite certain clean room facilities at the receiving end might be necessary and need to be ready.

[21] Candidate documents for this could be the configuration status accounting report (CSAR) or the software configuration file (SCF).

If a FAIL occurs at a test step to which the verification of a requirement has been traced, the verification of this requirement could not take place in this run. It is highly recommended to clearly define the criteria for a pass and fail and also any intermediate state[22] as part of the TRR, to avoid lengthy discussions during the test execution.

In case an anomaly in the system is observed it needs to be documented with a clear reference and description. The description needs to comprise all the details that will allow its reproduction later on-wards, which is a prerequisite for its correction. Depending on the nature of the anomaly and its impact it can be termed as a *software problem report* (SPR), *anomaly report*, or *non-conformance-report* (NCR). The term NCR is usually reserved for anomalies that have a wider impact (e.g., affecting functionality at system level) and are therefore subject to a standardised processing flow (see e.g., [32]).

It is highly recommended to hold a quick informal meeting at the end of every day throughout the campaign, in order to review the test results of each day and ensure proper documentation with a fresh memory. These daily reports are very useful for the generation of the overall test report at the end of the campaign, which is one of the main deliverable a the *test review board* (TRB). It is important to always keep in mind that test evidence is the main input to the verification control board (VCB) which is the place where the verification of a requirement is formally approved and accepted.

2.7 Summary

The objective of this chapter was to introduce the basic and fundamental principles of systems engineering (SE) that need to be applied in any project aiming to deliver a product that fulfils the end user's needs. SE activities must be closely coordinated with those of the project management domain and therefore the roles of the systems engineer and the project manager should be preferably occupied by good friends in order to guarantee project success. Especially at the project kick-off such a close interaction will favour the generation of a more realistic working schedule and enable the setup of more meaningful contractual hierarchies and collaborations, allowing to control the project cost and schedule more efficiently.

The importance to develop a good understanding of the existing risks in a project and the ability to properly categorise and continuously monitor them throughout the entire project life cycle has been briefly discussed. A risk register provides a good practice to document this but should be considered as a living document that

---

[22] A possible intermediate state is the so called qualified pass (QP) which can be used if a non expected behaviour or anomaly has occurred during the test execution which however does not degrade the actual functionality being tested.

requires continuous updates and reviews so it can help the project to take adequate risk mitigation measures in time.

The two basic life cycle models of sequential and incremental development have been explained in more detail. A profound understanding of the main characteristics (in terms of commonalities and differences) is an important prerequisite to be able to engage in the new lean and agile methods. The later ones can be considered as different flavours of the incremental approach[23] and have originally emerged in the software development domain but in recent years became more present at SE level. The modern lean and agile techniques should not be seen as practices that contradict the old proven methods but rather as a new way to apply SE principles in a more tailored and efficient manner with the sole intention to avoid rework at a later project stage.

Model based systems engineering (MBSE) has been briefly introduced as an approach to improve the generation of requirements (specification) by expressing them in executable models that are easier to visualise and to test. MBSE can improve the accuracy, consistency, and quality (e.g., less ambiguous) of a requirements baseline which in return improves the system design and implementation. At the same time the risk of to introduce change requests to correct or complement functionality at later project stages can be reduced significantly.

Finally some aspects of quality assurance (QA)[24] have been explained that affect both the development and verification stage of a project. In the development phase the application of software standards has been mentioned as an important means to reduce the likelihood of software bugs. The verification stage will benefit from an optimised planning and documentation of the test campaigns with clearly defined review milestones at the beginning and the end. It is worth to mention that the need and benefit of QA can sometimes be underestimated and easily considered as an additional burden to already high workloads (e.g., additional documentation that is considered unnecessary). Such a view should always be questioned as insufficient QA can lead to design or implementation errors. Inefficient test campaigns can lead to poor documentation and missed requirements in the verification process. In both cases additional and potentially even higher workload will be required and most likely imply schedule delays and cost overrun.

References

1. International Organization for Standardization. (2015). Systems and software engineering — System life cycle processes. ISO/IEC/IEEE 15288:2015(en)
2. International Council of Systems Engineering. (2015). *Systems Engineering Handbook*. A guide for system life cycle processes and activities. INCOSE-TP-2003-002-04.

---

[23] Some readers might strongly disagree with this statement but the author still dares to express this as his personal view.

[24] In some projects the term product assurance (PA) is used instead.

3. National Aeronautics and Space Administration. (2016). *NASA systems engineering handbook.* NASA SP-2016-6105, Rev.2.

4. European Cooperation for Space Standardization. (2017). System engineering general requirements. ECSS-E-ST-10C, Rev.1.

5. Belvoir, F. (1993). Committed life cycle cost against time. Belvoir: Defense Acquisition University.

6. Project Management Institute. (2021). *A guide to the project management body of knowledge* (7th edn.). ANSI/PMI 99-001-2021, ISBN: 978-1-62825-664-2.

7. Ruskin, A. M., & Eugene Estes, W. (1995). *What every engineer should know about Project Management* (2nd edn.). New York: Marcel Dekker. ISBN: 0-8247-8953-9.

8. European Cooperation for Space Standardization. (2009). Space engineering, Technical requirements specification. ECSS-E-ST-10-06C.

9. European Cooperation for Space Standardization. (2009). Space project management, project planning and implementation. ECSS-M-ST-10C.

10. European Cooperation for Space Standardization. (2008). Space management, Organization and conduct of reviews. ECSS-M-ST-10-01C.

11. Forsberg, K., & Mooz, H. (1991). The relationship of systems engineering to the project cycle. In *Proceedings of the National Council for Systems Engineering (INCOSE) Conference, Chattanooga*, pp. 57–65.

12. Forsberg, K., Mooz, H., & Cotterman, H. (2005). *Visualizing project management* (3rd edn.). Hoboken, NJ: Wiley.

13. European Space Agency. (1996). Ariane 5, Flight 501 Failure. Inquiry Board Report. Retrieved March 5, 2022 from https://esamultimedia.esa.int/docs/esa-x-1819eng.pdf

14. Douglas, B. P. (2016). *Agile systems engineering.* Burlington: Morgan Kaufmann. ISBN: 978-0-12-802120-0.

15. Scaled Agile Incorporation. (2022). Achieving business agility with SAFe 5.0. White Paper. Retrieved March 5, 2022 from https://www.scaledagileframework.com/safe-5-0-white-paper/

16. Womach, J. P., & Jones, D. T. (1996). *Lean thinking.* New York, NY: Simon & Shuster.

17. Toyota Motor Corporation. (2009). Just-in-time - Productivity improvement. Retrieved March 5, 2022 from www.toyota-global.com

18. Scaled Agile Incorporation. (2022). Agile teams. Retrieved March 5, 2022 from https://www.scaledagileframework.com/agile-teams/

19. Scaled Agile Incorporation. (2022). Agile release train. Retrieved March 5, 2022 from https://www.scaledagileframework.com/agile-release-train/

20. Schwaber, K., & Sutherland, J. (2022). The definitive guide to scrum: The rules of the game. Retrieved March 5, 2022 from https://scrumguides.org/docs/scrumguide/v2020/2020-Scrum-Guide-US.pdf

21. Beck, K. (2004). *Extreme programming explained: Embrace change* (2nd edn.). Boston: Addison-Wesley.

22. Lui, K. M., & Chan, K. (2006). Pair programming productivity: Novice-novice vs. expert-expert. *International Journal of Human-Computer Studies, 64*, 915–925.

23. Scaled Agile Incorporation. (2022). System team. Retrieved March 5, 2022 from https://www.scaledagileframework.com/system-team/

24. Scaled Agile Incorporation. (2022). Team Kanban. Retrieved March 5, 2022 from https://www.scaledagileframework.com/team-kanban/

25. Object Management Group®. (2019). OMB systems modeling language (OMG SysML®), Version 1.6. Doc Number: formal/19-11-01. Retrieved March 5, 2022 from https://www.omg.org/spec/SysML/1.6/

26. Object Management Group®. (2022). Official website. Retrieved March 5, 2022 from https://www.omg.org/

27. European Cooperation for Space Standardization. (2018). Space product assurance: Quality assurance, Rev. 2. ECSS-Q-ST-20C.

28. Radio Technical Commission for Aeronautics (RTCA). (1992). Software considerations in airborne systems and equipment certification. Doc DO-178B.

29. Radio Technical Commission for Aeronautics (RTCA). (2011). Software considerations in airborne systems and equipment certification. Document DO-178C.
30. Youn, W. K., Hong, S. B., Oh, K. R., & Ahn, O. S. (2015). Software certification of safety-critical avionic systems and its impacts. *IEEE A&E Systems Magazine, 30*(4), 4–13. https://doi.org/10.1109/MAES.2014.140109.
31. Galileo Project Office. (2016). Galileo software standard for ground (GSWS-G). GAL-REQ-ESA-GCS-X/0002, Issue 1.0.
32. European Cooperation for Space Standardization. (2018). Space product assurance, nonconformance control system. ECSS-Q-ST-10-09C.

Chapter 3
The Space Segment

The purpose of ground segment engineering is to design, develop, and deploy an infrastructure that is able to operate a space vehicle throughout its entire life span. As the space segment can strongly influence the design of the ground segment, it is obvious that the ground segment engineer must develop a profound understanding of the space vehicle's architecture, its various subsystems, and how the ground to space interaction need to be performed throughout the various stages of its life cycle. Whenever the term *space segment* is mentioned here, it has to be understood in a broader context, as it can represent a single satellite, a satellite constellation (or formation), an interplanetary probe, a manned space transportation system or even an entire space station. Space segment systems engineering is a complex subject matter that fills a large number of excellent text books available in the open literature. Only a high level overview will be provided here and the reader is encouraged to consult the cited references in case more information is needed.

The interaction between the ground and space segments can be better described via the definition of a set of *ground-to-space interface points* which are shown in Fig. 3.1 as grey boxes and define the areas where exchange of information between the two segments occurs and which specific satellite subsystem are involved. Due to the high relevance of these interface points they need to be clearly defined by the ground segment engineer both in terms of the expected data contents as well as its completeness and correctness. Any inaccurate, incomplete, or late delivery of the required information can potentially imply delays in the ground segment readiness which in a worst case will even cause a launch delay.

3.1 System Design

The fundamental architecture of a satellite is mainly driven by the payload it operates and the related performance requirements it has to fulfil in order to meet its mission objectives. These can be very different and project dependant. A satellite

B. Nejad, *Introduction to Satellite Ground Segment Systems Engineering*, Space Technology Library 41, https://doi.org/10.1007/978-3-031-15900-8_3

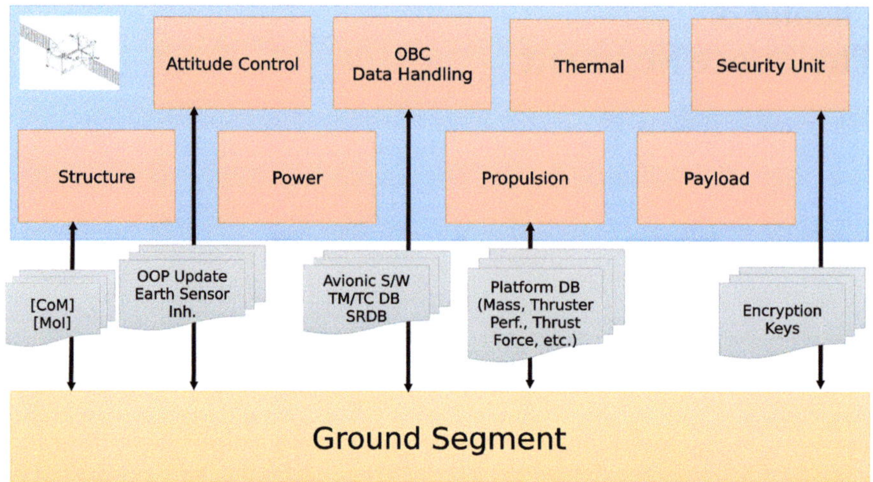

Fig. 3.1 Overview of main satellite subsystems and their respective interface points to the ground segment (refer to text). SRDB = Satellite Reference Database, [MoI] = Moments of Inertia matrix, [CoM] = Center of Mass interpolation table, OOP = Onboard Orbit Propagator

platform that carries a scientific remote sensing payload (e.g., a space telescope) will usually have to meet quite stringent requirements on its pointing accuracy and stability (jitter) which will have an impact on the design of its attitude control system. A telecommunication satellite in contrast usually has a lower demand for accurate pointing but a much higher one on the power subsystem to feed its transponder so it can meet the required RF transmission characteristics and the link budget. Interplanetary spacecraft that travel to the inner Solar System have to face very high thermal loads whereas a voyage to the Jovian system implies extreme radiation doses that requires extensive radiation shielding to protect the satellite's electronic circuits and ensure it's survival. With this in mind one can state that satellite platforms will only resemble in projects with similar mission objectives, payloads, and operational environments. Independent of this one can however find a similar set of standard subsystems in all satellite platforms which are briefly summarised below.

3.1.1   Propulsion

The propulsion system enables the satellite to exert a thrust and change its linear momentum. This is needed to insert a satellite into its operational orbit position and to perform any correction of its orbital elements during its nominal lifetime (station keeping). Propulsion systems can be categorised into chemical, nuclear, or electrical systems. Nuclear based system have never reached a wider deployment in space and will therefore not be discussed in more detail here.

Chemical propulsion can be either realised as monopropellant or bipropellant systems, where for the latter one two propellant components (fuel and oxidiser) are needed. The advantage of bipropellant systems is a higher specific impulse which however comes at the cost of additional complexity in the design due to the need of separate propellant tanks for each component and additional pumping and feed devices from the tanks to the injector.

Solid propellant systems have a much simpler design but can provide very high trust values and are therefore often used to support the early stages of launch vehicles. Oxidiser and fuel are stored in the correct ratio in the form of a solid inside the combustion chamber. This avoids the need to implement separate tanks for oxidiser and fuel and any transport mechanism to flow these into the combustion chamber (see e.g., [1]). An important limitation of solid propellant systems however is that once they have been ignited, the combustion process proceeds until all of the contained propellant is consumed.

The propulsion system deployed in satellites is usually a monopropellant system using Hydrazine (N_2H_4) which decomposes in an exothermic reaction to the nitrogen, ammonia, and hydrogen and provides a specific impulse in the range between 200–250 s (cf.[2]). With a freezing point at 275 K (i.e., 1.85 C°) and a boiling point at 387 K (i.e., 113.85 C°) the storage of liquid Hydrazine is technically not very demanding and the propellant is usually kept in tanks under pressure of an inert gas such as nitrogen or helium. If a thruster is activated, a latch valve opens and the Hydrazine flows to a heated metal grid called cat-bed heater, where the decomposition is enhanced.

Another monopropellant type system is the cold gas thruster which has a very low thrust and is therefore typically used to support the attitude control of satellite's with a high demand for pointing accuracy and stability (e.g., space telescopes). The propellant used is an inert gas (e.g., nitrogen, argon, propane) which is kept under high pressure. Upon activation of a nozzle, the gas is fed to a number of small thrusters where it expands and provides the small impulse bits required by the attitude control system.

Electrical propulsion systems are usually limited to much lower thrust levels but have significant advantages in terms of propellant efficiency and system design. Due to their low thrust capability they are not suitable for high power operations like a launch or the apogee raising manoeuvres needed to circularise an orbit. More recently satellite projects are increasingly implementing this technology with one prominent example being the ESA/Jaxa Bep-Colombo mission to Mercury which uses solar electrical propulsion for its interplanetary cruise trajectory (cf.,[3]).

3.1.2   Attitude Control

The attitude control system (ACS) enables a satellite to fulfil its pointing requirements which are defined in terms of pointing orientation, accuracy, and stability. The orientation will be primarily dictated by the payload and the mission objectives.

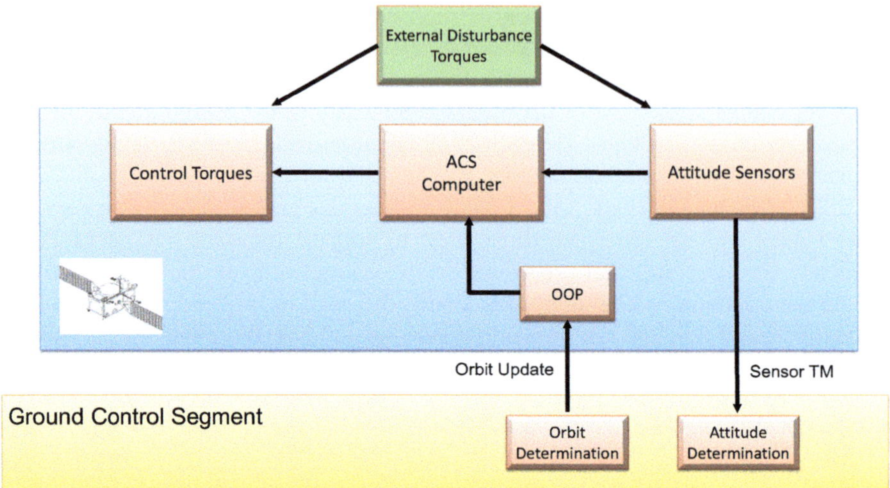

Fig. 3.2 Overview of the satellite attitude control system's (ACS) main components. OOP = Onboard Orbit Propagator (see Sect. 6.6)

Remote sensing devices, TV, and telecommunication broadcasting antennas need to point to the Earth whereas a space telescope needs to be flexible and able to point to any location in space in order to target a star, planet, or any other deep space object. Additional pointing constraints will stem from other subsystem like solar arrays having a need to point to the Sun or thermal radiators with the need to be oriented versus cold space.

From an architecture point of view, the ACS system can be decomposed into three subsystem groups: the sensors able to measure the current attitude, the actuators to exercise the necessary control torque to achieve a target attitude, and the computer that runs the algorithm to determine the current attitude and the required torque levels to reach the targeted one (see Fig. 3.2). These type of algorithms are based on the mathematical formulation of attitude dynamics which is quite a complex subject matter that for the sake of brevity will not be elaborated in more detail here and the interested reader is encouraged to consult the broad range of existing specialised literature on this topic (see e.g., [4, 5]). As any attitude determination algorithms will need to perform a number of coordinate transformations (e.g., from inertial to body fixed satellite frame), the satellite will require an estimate of its orbit position and velocity which it receives from its onboard orbit propagator (OOP) subsystem that is designed to predict the satellite's state vector at any required epoch t. As the OOP vectors are usually derived by propagation from an initial state vector the OOP output accuracy will decrease with time and will need to receive regular updates from ground. This regular OOP update is an important ground support function and is therefore described in more detail in Sect. 6.6.

Independent of the actual satellite design its attitude control will always be based on the basic physical principles to manage the satellite's angular momentum $\vec{H}$.

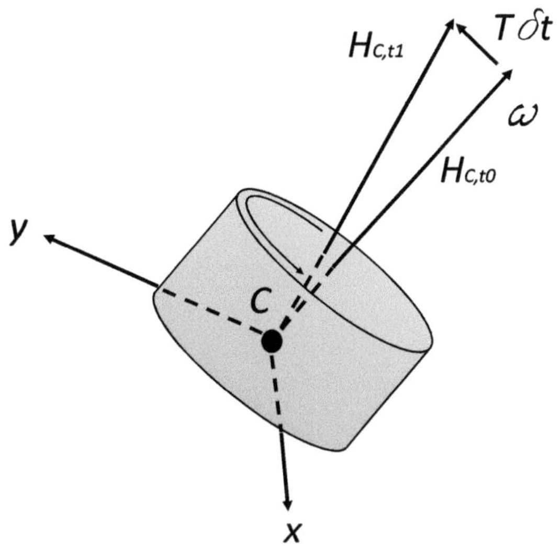

Fig. 3.3 Rotating axis-symmetric spinning body with angular momentum vector $\vec{H}_C$ and rotation vector $\vec{\omega}$ along the same direction. An external torque acting over a time span $T\,\delta t$ is able to change magnitude and direction of the angular momentum vector $\delta \vec{H}_C$

This is a vector quantity that can be derived by summing up the moments of the momenta of all particles at their location $\vec{r}$ inside the rigid body.

$$\vec{H}_O = \sum (\vec{r} \times m\,\vec{v}) \tag{3.1}$$

The subscript '$_O$' in Eq. (3.1) defines the common origin or reference point from which $\vec{r}$ and $\vec{v}$ of each mass particle is defined. The angular momentum vector $\vec{H}$ has in the context of rotational motion an equivalent meaning as the linear momentum vector $\vec{L} = m \times \vec{v}$ for translational movement. An external torque $\vec{T}$ has, equivalent to a force in linear motion, the ability to change both magnitude and direction of the angular momentum vector, i.e., $\delta \vec{H}_C = \vec{T}\,\delta t$. A torque component parallel to $\vec{H}$ will only modify its magnitude, whereas an orthogonal torque component T_N will change the direction of $\vec{H}$ into the same direction as $\vec{T}$ which causes the well known *precession* of a gyroscope (see Fig. 3.3). It is worth to remember the following basic properties of the angular momentum vector $\vec{H}$:

- In contrast to external torques, no change of angular moment can be achieved via internal torques (e.g., fuel movement or movement of mechanisms inside the satellite).
- A flying satellite is always subject to natural external disturbance torques (e.g., solar radiation, gravity gradient, aerodynamic), which will imply a progressive build up of angular momentum over the satellite's lifetime. Therefore the satellite designer will need to fit the satellite with external torquers (e.g., thrusters or magnetic coils) in order to control (or dump) this momentum build up.
- Activation of thrusters that are not precisely pointing through the centre-of-mass can change the orientation of the angular momentum vector and destabilise its attitude. To avoid this unwanted circumstance, satellites are sometimes

artificially spun up prior to large manoeuvres. This increase the magnitude of $\vec{H}$ and for a given thrust T reduces the $d\vec{H}/dt$ term. This is referred to as building up *gyroscopic rigidity* in space.

For rigid bodies it is convenient to choose its centre-of-mass (CoM) as a reference for the expression of the angular momentum vector as it can then be expressed as

$$\vec{H}_C = [I_C]\vec{\omega} \tag{3.2}$$

where the subscript $_C$ refers to the CoM, $[I_C]$ is the inertia tensor, and $\vec{\omega}$ is the angular velocity vector relative to an inertial reference frame. The inertia tensor in matrix notation is defined by

$$[I_C] = \begin{bmatrix} I_{xx} & -I_{xy} & -I_{zx} \\ I_{xy} & -I_{yy} & -I_{yz} \\ I_{zx} & -I_{yz} & -I_{zz} \end{bmatrix} \tag{3.3}$$

where I_{xx}, I_{yy}, and I_{zz} are the moments of inertia, and I_{xy}, I_{yz}, and I_{zx} the products of inertia. The later are a measure of the mass symmetry of the rigid body and are zero if the rotation axis is chosen to be one of the body's *principal axis* which represent the *Eigenvectors* of the inertia matrix. For a spacecraft that contains moving parts like momentum or reaction wheels, the additional angular momentum of each wheel, e.g., $\vec{H}_{wh} = I_{wh}\vec{\omega}_{wh}$ needs to be added to the overall angular momentum yielding:

$$\vec{H}_C = \begin{bmatrix} I_{xx}\omega_x - I_{xy}\omega_y - I_{zx}\omega_z + H_{x,wh} \\ I_{yy}\omega_y - I_{yz}\omega_z - I_{xy}\omega_x + H_{y,wh} \\ I_{zz}\omega_z - I_{zx}\omega_x - I_{yz}\omega_y + H_{z,wh} \end{bmatrix} \tag{3.4}$$

Depending on the overall attitude motion, one can distinguish between spinning and three-axis-stabilized satellites. Whereas the first one will always have a momentum bias both options are possible for the latter one. The main advantage of momentum bias is to gain gyroscopic rigidity which can be either applied temporarily (e.g., during rocket firing) or permanently as done for spinners. For the latter case, the rotation axis will usually be chosen in a direction that does not have to change during satellite operations. A spinning spacecraft is schematically depicted in Fig. 3.3 for which the rotation axis $\vec{\omega}$ is aligned with the angular momentum axis $\vec{H}$. This alignment will only be stable if the rotation axis is a principal axis (i.e., an *Eigenvector* of the bodies' inertia matrix).

Attitude determination depends on the input provided by specialised sensors of which the most common ones are summarised in the upper part of Fig. 3.4. An Earth sensor is designed to detect the Earth's horizon and is usually operated in the infrared spectrum where additional optimisation effects can be exploited (e.g., less variation between peak radiation, no terminator line, presence of the Earth in IR even during eclipses). *Static Earth-horizon* sensors are usually used for high altitude

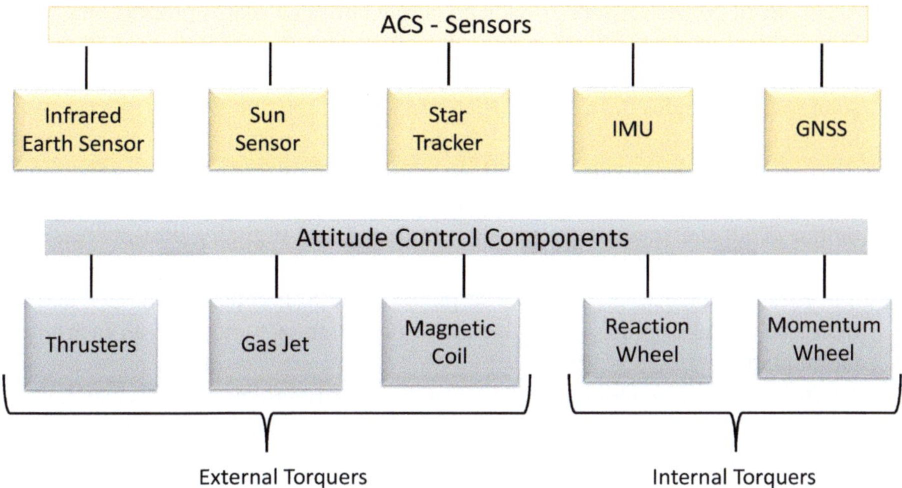

Fig. 3.4 Overview of ACS sensors and torquers

orbits (MEO and GEO) for which the apparent radius of the Earth[1] is small enough to locate transition detectors in a static arrangements that ensures their overlap with the border of the apparent Earth disc. This will allow the sensor to measure a displacement of the detectors and derive a deviation of roll and pitch angles from a reference. Due to the circular shape of the Earth the measurement of the yaw angle (rotation around the local vertical) is however not possible with that same device.

As the apparent Earth diameter is extensively larger for a LEO type orbit[2] *scanning Earth-horizon sensors* are used to determine the nadir direction. As implied by their name such units implement a scanning mechanism that moves a detector with a small aperture (pencil beam) on a conical scan path. When crossing the horizon the sensor generates a square-wave shaped pulse which is compared to a reference pulse and its offset used as a measure of the satellite's deviation from that reference attitude (see e.g. [6] for more details).

Sun Sensors are widely used on satellites due to the advantage that the Sun's angular radius is fairly insensitive to the chosen orbit geometry (i.e., 0.267 deg at 1 AU) and allows it to be used as a point source approximation in the sensor software. Furthermore, the Sun's brightness makes the use of simple equipment with minimal power requirement possible. There is a wide range of designs which differ both in their field of view (FOV) and resolution specification and lead to the terms Fine Sun Sensor (FSS) and Coarse Sun Sensor (CSS). Based on their basic design principle the following main categories are distinguished (cf., e.g. [4]):

---

[1] The Earth apparent radius at an orbit altitude h is given by $\alpha_{R_E} = 2 \ \arcsin[R_E/(R_E + h)]$, which yields around 17.5 deg at GEO altitude.

[2] As an example, at an orbit altitude of 500 km the apparent Earth has a diameter of ca. 135 deg.

- *Analogue* or *cosine* sensors provide an output current I that is proportional to the Sun incident angle Θ on the photocell normal vector, i.e., $I(\theta) = I(0)\cos(\theta)$. An important limitation of this sensor type is a decrease in accuracy at increasing solar incident angles θ. In order to determine the three Sun vector components in the satellite's body frame, two sensor units oriented perpendicular to each other (two-axis Sun sensing) are required. A combination of several devices distributed over the satellite's entire body surface provides an omnidirectional FOV. The output of all the sensors are usually combined and digitally post-processed to derive the Sun vector.
- *Sun presence detectors* provide a response if the Sun enters their FOV and are therefore useful to protect sensitive instrumentation or to position an instrumentation platform.
- *Digital sensors* provide a higher accuracy compared to analogue sensors, especially for high deviations of the Sun vector from the optical axis with the drawback of higher cost and complexity.

Star sensors or *trackers* belong to the most accurate attitude sensors able to achieve sub-arc-second type measurements. They only require visibility to stars and can therefore also be used for interplanetary voyages. A star sensors is a quite complex device that needs to implementing software to store and access star charts and techniques to distinguish between objects. The number of stars that can be processed by an algorithm will determine the achievable accuracy.

Inertial Measurement Units or IMUs are used to measure changes in the satellite's attitude by measuring the three components of the satellite's angular velocity vector, i.e., ω_x, ω_y, and ω_z. The determination of the satellite's attitude at a required time t requires the knowledge of an initial value from a reference sensor which is referred to as a calibration point. In between two calibration points the IMU can integrate the measured angular rates and provide an attitude profile who's accuracy will decrease with time. The spacecraft attitude provided by an IMU is of special importance during times when no attitude information from other sensors is readily available. It is therefore a key instrument for critical satellite operations during eclipse periods.

A complete different type of attitude determination is based on the use of navigation signals (e.g., GPS, Galileo, Glonass) taking advantage of already deployed GNSS receivers the satellite uses to obtain its orbit position. The navigation signal based attitude measurement concept uses a set of GNSS patch antennas which are mounted at different platform locations as illustrated in Fig. 3.5. The distance between these antennas is referred to as the measurement baseline L and is accurately measured before launch. If the same GNSS satellite is seen by both patch antennas, the two measured pseudoranges will differ by $\delta\rho = L\cos\Theta$. Available observables to determine $\delta\rho$ are the code carrier phase measurements and the Doppler shifts of the carrier frequency. This method has already been demonstrated in orbit at several occasions and attitude determination accuracies down to 0.1 deg could be demonstrated (cf., [7]).

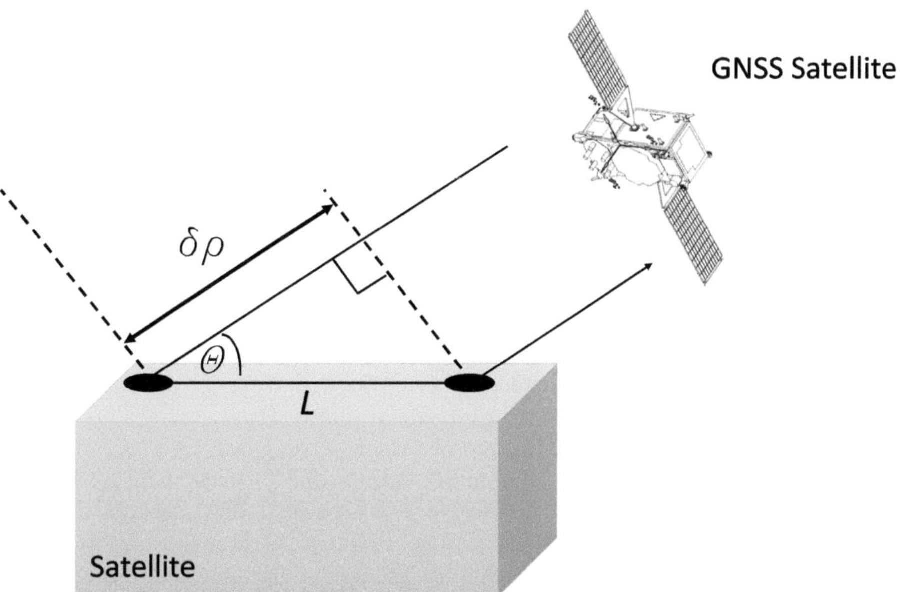

Fig. 3.5 Principle of a GNSS derived attitude measurements using two patch antennas mounted on a satellite surface at distance L

Whereas the ACS sensors explained above are used to determine the satellite's attitude, actuators are used to change it, either to correct the current attitude or to achieve a new one. Actuators can be categorised into two groups, *external torquers* affecting the total angular momentum (and necessary to counteract external torques) and *internal torquers* that only redistribute the momentum between its moving parts (see lower panel of Fig. 3.4).

Thrusters are usually mounted in clusters with different directions and are mainly applied for orbit corrections but they can also assist the attitude control system. Thrusters have the advantage to provide altitude independent torque magnitudes (as opposed to magnetic torquers) but come with the drawback that their accumulated thrust and exercised torque is determined via their switch-on duration. If thrusters are used as the prime attitude control means of a satellite, a high level of activation cycles might be needed which will reduce the lifetime of the entire subsystem and with this even that of the satellite. Reaction and momentum wheels are therefore the preferred choice and thrusters serve as momentum dumping devices.

Cold Gas systems are a specific type of thruster designed for small thrust levels, typically in the order of 10 mN. Their design implements an inert gas (e.g., Nitrogen, Argon, or Freon) which is stored at high pressure and released into a nozzle on activation where it simply expands without any combustion process involved. The thrust force stems from the pressure release of the inert gas into space. The low thrust level and ability to be operate with very short switch-on times allows the achievement of small impulse bits in the order of only 10E-04 Ns which is needed to achieve high-precision pointing requirements.

Magnetic torquers are rod like electromagnets whose generated magnetic field depends on the current I that flows through them. The resulting torque T stems from the interaction with the local Earth magnetic field vector $\vec{B}$ via the following relation

$$\vec{T} = \frac{n\,I\,A}{\vec{e} \times \vec{B}} \tag{3.5}$$

where n is the number of wire turns on the coil, A the cross-sectional area of the magnetic coil, and $\vec{e}$ the unit vector of the coil's axis.

Reaction Wheels (RWs) and *Momentum Wheels* (MWs) are precision wheels that rotate about a fixed axis with an integrated motor. They need to implement a reliable bearing unit which despite high rotation rates can guarantee a lifetime that is commensurate with the overall satellite life time (usually 15 years). The difference between RW and MW is mainly the way they are operated. RWs are kept at zero or very low rotation and through their mounting in orthogonal direction used to provide three-axis control of the satellite. MWs are operated at high rotation rates (usually in the range between 5000 to 10000 r.p.m) and used to provide a momentum bias to the satellite.

3.1.3   Transceiver

The satellite transceiver subsystem has to establish the RF link to the ground segment's TT&C station network (refer to Sect. 5). The main components are depicted in Fig. 3.6 and are briefly described here. The *diplexer* separates the transmit and receive path and comprises low-pass and band-pass filters in order to prevent that out-of band signals reach the receiver's front end low noise amplifiers (LNAs). The received signal is routed to the receiver front-end (RX-FE) were it is amplified and down-converted to the *intermediate frequency* (IF) of 70 MHz. This conversion requires the transceiver to accommodate its own frequency generator which usually is a temperature controlled oscillator (TCXO). The IF signal then enters the *modem interface unit* (MIU), where it is converted into the digital domain for further signal processing (bit synchronization, demodulation, and modulation).

On the transmit chain the TM frames are modulated by the IMU onto the baseband signal and then transformed from the digital to the analogue domain. Following up-conversion (UC) to high-frequency the signal is amplified to the required up-link power level by the *high power amplifier* (HPA) unit. The adequate amplification level depends on the *link budget* which needs to consider an entire range of parameters in the overall transmission chain. These comprise the antenna gain, the propagation distance, atmospheric losses, receiver noise level, transmitter and receiver losses (refer to Sect. 5.5). After amplification the signal is routed to the Diplexer for transmission to ground.

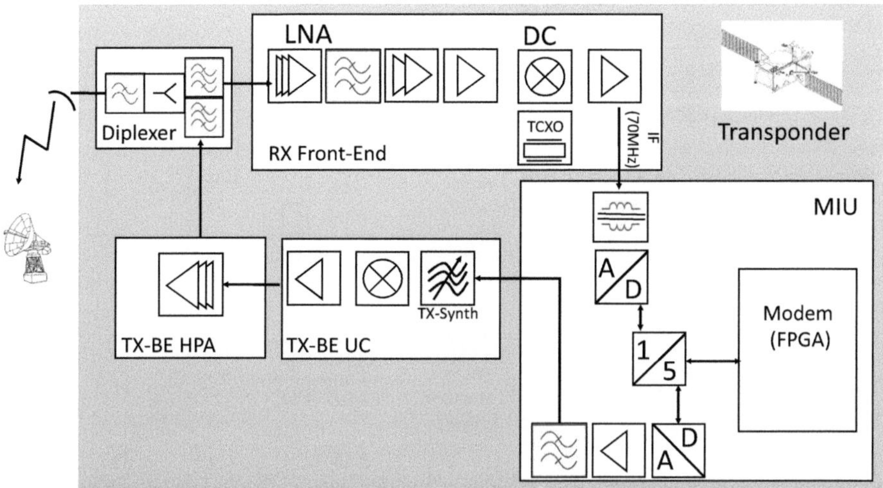

Fig. 3.6 Main components of a typical satellite transceiver. LNA = Low Noise Amplifier, DC = Down Converter, MIU = Modem Interface Unit, TX-BE SSPA = Transmitter Backend Solid State Power Amplifier

The overview provided here is quite generic and the main architectural differences among various transceiver models will usually differ at the level of the MIU (e.g., need to support different modulation schemes) and the specification of the HPA.[3]

3.1.4 Onboard Computer and Data Handling

The main components of a satellite onboard computer (OBC) and its interfaces to other satellite subsystems are shown in Fig. 3.7. The core component of the OBC are the two redundant central processing units (CPUs) which provide the necessary processing power and memory to run the onboard software (OBSW) and to process all the incoming data from the satellite's subsystems. In contrast to computer systems deployed on ground, CPUs in space need to withstand much harsher environmental conditions which comprise high energy radiation environments (e.g., the Earth's van Allen belt or the Jovian system), extreme thermal conditions (e.g., missions to the inner part of the Solar system), aggressive chemical environment (e.g., atomic oxygen of Earth's outer atmosphere), or launcher induced vibrations [8]. Due to the critical importance of the CPU for the overall functioning of the

---

[3] The required amplification level will drive the energy consumption of the satellite and mainly depend on the orbit altitude (propagation path) and the required signal level on ground.

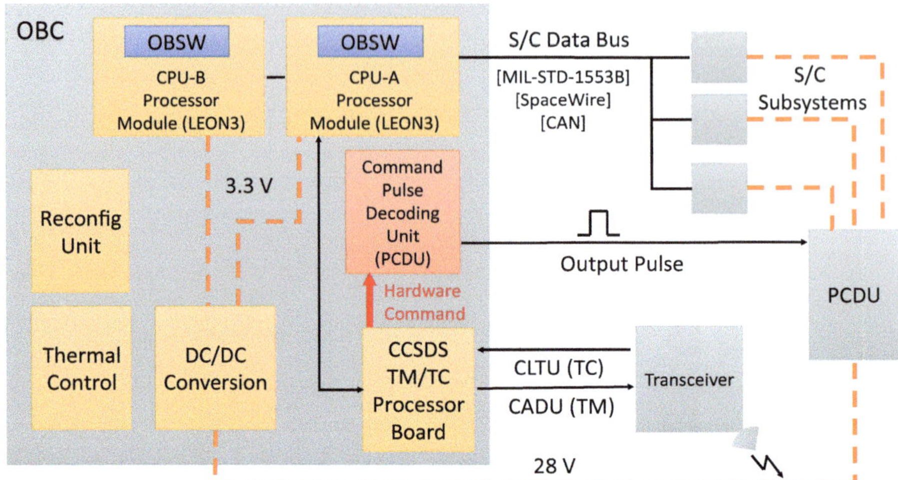

Fig. 3.7 Satellite onboard computer (OBC) components. Solid lines represent data interfaces whereas dashed lines power connections. OBSW = Onboard Software, PCDU = Power Control and Distribution Unit, CLTU = Command Link Transfer Unit (CCSDS TC format), CADU = Channel Access Data Unit, CAN = Controller Area Network

OBC, redundant processor architectures are the standard design. The interface with other satellite subsystems is realised via the so called *spacecraft bus* which can be based on a variety of existing standards. The most frequent one in use is the US military standard MIL-STD-1553B [9] which consists of a *bus controller* able to serially connect to up to 31 "Remote Terminals". Alternative standards in use are *SpaceWire* defined by ESA ECSS [10] or the *Controller Area Network* (CAN) interface (cf.[11]) which was originally developed by the automotive industry and has only later been adapted for the use in other areas including space applications.

The communication of the satellite to the ground segment's TT&C station is mainly performed using the CCSDS standardized *Space Link Extension* or SLE protocol which is described in more detail in Sect. 7.2.1. This standard describes the layout of packets and segmentation of these packets into frames and higher order structures to build the *Command Link Transfer Units* (CLTU) for TC uplink and the *Channel Access Data Units* (CADU) for TM downlink. The TC decoding and TM encoding is performed by a dedicated CCSDS TM/TC processor board which interfaces to the modem of the satellite transceiver subsystem. These boards are usually entirely hardware based (ASICs or FPGAs) and by design explicitly avoid the need of a higher level software which would need to be implemented as part of the OBSW. This has the important advantage that the CCSDS processor board can still interpret emergency commands from ground in case the OBSW is down. Such emergency commands are referred to as *high priority commands* (HPCs) and are identified by a dedicated command packet field code. This allows them to be immediately identified by the CCSDS processor board which can then route them directly to the *command pulse decoding unit* (PCDU) and bypass the OBSW. Following the directives in the HPC, the PCDU sends analogue pulses over *pulse lines* with different lengths in order to steer the *power control and distribution unit* (PCDU). Possible scenarios could be the emergency deactivation of certain power consumers or a power cycle (reboot) of the entire OBC.

In case of a component failure, the *reconfiguration unit* has to initiate the necessary "rewiring" of a redundant component (e.g., the main processor boards) with a non redundant unit (e.g., CCSDS TM/TC board) inside the OBC.[4] The switchover can either be commanded from ground or initiated via the satellite internal FDIR sequence. Furthermore, the reconfiguration unit needs to track the health status of each OBC component and update this state in case it is reconfigured. This information is kept in a dedicated memory location referred to as *spacecraft configuration vector* and any switchover activity is communicated to ground via dedicated *high priority telemetry* (HPTM).

Another important OBC component is the DC/DC converter which converts the satellite bus standard voltage supply of 28 V to the much lower 3.3 V that is usually required by the various OBC components.

From a mechanical design point of view, OBCs are usually composed of a set of printed circuit boards, with each of them mounted in a dedicated aluminium frame. These frames are then accommodated into an overall OBC housing. As every one of the these circuit boards dissipates heat they are monitored via thermistors located on the board surfaces which is shown via the *Thermal Control* box in Fig. 3.7.

3.1.5 Power

The power system has to ensure that sufficient electrical power is generated, stored, and distributed to the consumers so they are able to properly function. The power subsystem is usually structured into a primary and secondary block and a *power control and distribution unit* (PCDU) which is schematically shown in Fig. 3.8. The primary block is responsible to generate power using solar arrays, fuel cells, or nuclear generators referred to as *Radioisotope Thermoelectric Generator* or RTGs.[5] The secondary block is responsible for the storage of energy and therefore comprises a set of batteries. This storage capacity needs to be correctly dimensioned and is based on the outcome of a dedicated analysis taking into account the maximum power demand (load) of all the satellite's subsystems, the available times the satellite is exposed to sunlight, and the maximum duration of encountered eclipses. Another important aspect for the power budget is the solar array degradation rate due to

---

[4] A reconfiguration of an OBC external component like a redundant thruster branche is usually performed by the OBSW.

[5] RTGs are generally deployed in deep-space missions targeting the outer planets of the Solar System where the power received from the Sun is too week to satisfy the consumer's needs.

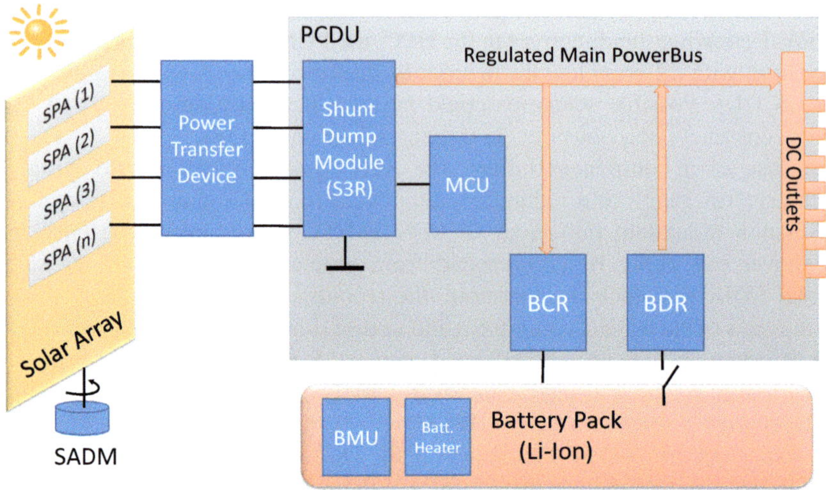

Fig. 3.8 Simplified view of a satellite's Electrical Power Subsystem (EPS). SPA = Solar Power Assembly, S3R = Sequential Switching Shunt Regulation, MCU = Mode Control Unit, BCR = Battery Charge Regulator, BDR = Battery Discharge Regulator, BMU = Battery Monitoring Unit

radiation which implies a reduction of generated power with ageing of the array, reaching its lowest value at the end of the satellite's lifetime.

As both the generated and consumed power levels usually vary during an orbit, the solar array segments (also referred to as *solar power assembly* or SPA) can be activated or deactivated in a staggered way. This is done by the *shunt dump module* (SDM) which steers the grounding of each SPA as shown in Fig. 3.8. The voltage sensing is done by the *mode control unit* or MCU.

The secondary block comprises the battery pack, a heater, and the *battery management unit*. The latter one monitors the battery temperature, the cell voltage, and the cell pressure. The *battery charge regulator* (BCR) ensure a constant current charge of the battery during sunlight operations and the *battery discharge regulator* (BDR) a constant supply to the main power electrical bus during eclipse times. Two different types of direct current power bus systems can usually be found in satellites, a regulated and an unregulated bus type. The regulated one provides only one single voltage (e.g., 28 or 50 V) whereas an unregulated one an entire range of voltage values. The satellite's subsystems (consumers) are connected to the outlets of the PCDU.

3.1.6 Thermal Control

The prime task of a satellite's thermal control system is to monitor and regulate the environmental temperature conditions of the entire satellite platform. Electronic components are the most sensitive parts onboard a satellite, and are nominally qualified to operate within a very narrow range around room temperature. Due to extreme temperature gradients in space, the thermal control system must implement active and passive measures to limit these in order to avoid excessive thermal expansion of the satellite structure which could lead to distortions or even structural instabilities.

Due to the absence of an atmosphere in space, radiation and conductance are the dominant heat transfer mechanisms and convection can be neglected. The main heat sources are externals ones that stem from solar radiation reaching the satellite either directly or through reflection and thermal infrared radiation from nearby planets. There is also an internal source which is the heat generated by the various electronic devices or batteries. The satellite can dissipate heat through its radiators that point to cold space. Thermal equilibrium is therefore reached when the amount of heat received from outside and any thermal dissipation inside the spacecraft equals the amount of energy that is radiated to deep space. This thermal balance can be expressed as (cf., Sec. 11 of [2])

$$A_\alpha\, \alpha\, J_{incident} = A_\epsilon\, \epsilon\, \sigma\, T^4 \tag{3.6}$$

where the left part of the equation represents the absorbed energy that is determined by the absorbing area A_α, the incident radiation intensity $J_{incident}$ (W/m), and the surface radiation absorbance α and the right part represents the radiated energy to space with the emitting surface A_ϵ, the emittance ϵ, and the equilibrium temperature of the satellite body T. If a spacecraft would behave like a perfect black body, the coefficients α and ϵ would be one. As both coefficients are material constants, the satellite manufacturer can influence the equilibrium temperature through the proper choice of surface materials based on their thermal coefficients which are known and readily available from open literature (see, e.g., [12] or [13]).

A detailed thermal design of a satellite cannot be purely based on the thermal equilibrium equation and a more refined formulation is required that also takes into account the thermodynamics inside the satellite specific structure. This formulation is referred to as the *thermal mathematical model* or TMM and needs to be formulated for each satellite during its design phase. The fundamental concept of a TMM is to discretise the satellite body into a number n of isothermal nodes as shown in Fig. 3.9. Each node is characterised by its own temperature T_i, thermal capacity C_i, and radiative and conductive heat exchange mechanisms to its surrounding nodes. If a node is located at the satellite's surface, dissipation to outer space also needs to be considered. The conductive heat transfer between node i and j can therefore be written as (cf. [14])

$$Qc_{ij} = h_{ij}\, (T_i - T_j) \tag{3.7}$$

where h_{ij} is the effective conductance between nodes i and j with their respective temperatures T_i and T_j. The term h_{ij} represents the effective thermal conductance

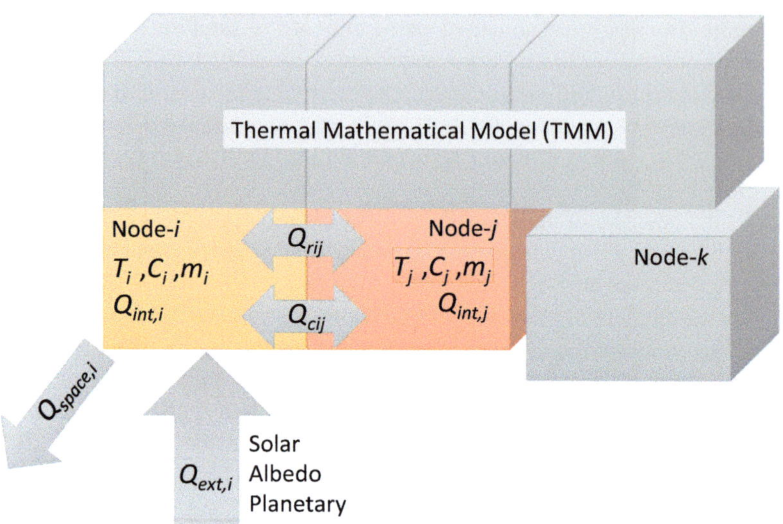

Fig. 3.9 The thermal mathematical model of a satellite showing the heat exchange of a single node i

between the two nodes. The radiative heat transfer between nodes i and j is given by

$$Qr_{ij} = A_i \, F_{ij} \, \epsilon_{ij} \, \sigma \, (T_i^4 - T_j^4) \tag{3.8}$$

where A_i is the emitting surface, ϵ_{ij} the effective emittance. The term F_{ij} is the so called *radiative view factor* that quantifies the fraction of the radiation leaving one surface and intercepted by the other.[6]

With the aforementioned definitions in place, the TMM can now be formulated as a set of n non-linear differential equations where each equation defines the thermodynamics for one single node i.

$$m_i \, C_i \, \frac{dT_i}{dt} = Q_{ext,i} + Q_{int,i} - Q_{space,i} - \sum_{j=1}^{n} Qc_{ij} - \sum_{j=1}^{n} Qr_{ij}. \tag{3.9}$$

Q_{int} is the internal heat dissipation of the node, Q_{space} the heat radiated to space given by

$$Q_{space,i} = \sigma \, \epsilon_i \, A_{space,i} \, T_i^4 \tag{3.10}$$

---

[6] For a more detailed treatment of the view factor, the reader is referred to more specialised literature like e.g., [14].

where $A_{space,i}$ is the surface area of node i facing cold space. Q_{ext} is the external heat input comprising contributions from the Sun, the Earth's albedo, and any visible planet according to

$$Q_{ext,i} = J_s\, \alpha_i\, A_{solar,i} + J_a\, \alpha_i\, A_{albedo,i} + J_p\, \epsilon_i\, A_{planetary,i} \qquad (3.11)$$

J defines the flux densities (W/m^2) of each heat source. The terms Qc_{ij} and Qr_{ij} represent the thermal contributions of heat conduction and radiation as defined in Eqs. (3.8) and (3.7) respectively. With the TMM in place to accurately model the thermal dynamics inside satellite the designer can now take appropriate measures to influence it with the means of passive or active thermal control techniques.

Examples of passive techniques comprise the already mentioned selection of surface properties that will influence the α/ϵ ratio in Eq. (3.6) or the laying of conducting paths inside the satellite structure which influence the heat conduction term Qc_{ij} defined in Eq. (3.7). More elaborate and widely used passive technologies are two-phase heat transport systems like *heat pipes*, *loop heat pipes*, or *capillary-pumped loops* which implement sealed tubes that contain a working liquid which is kept in thermodynamic equilibrium with its vapour. A porous structure ensures that the liquid and the vapour remain separated but can still interact. The exposure of the pipe to high temperature on one side (incoming heat) will lead to evaporation of the liquid which causes a pressure gradient between the two sides of the pipe resulting in a transport of vapour to the opposite side. Once the vapour arrives at the cooler temperature end, it condensates and releases heat there. Another much simpler passive technique is the use of insulation materials designed to minimise radiative heat exchange in vacuum. The most prominent example is the multi-layer insulation (MLI) blanket used as radiation shields for satellite external surfaces that are exposed to the Sun.

Active control techniques comprise heaters,[7] variable conductance heat pipes, liquid loops, shutters, or heat pumps. For a more detailed description of these devices, the reader is referred to more specialised literature (e.g., [14] or [15]).

3.2 Spacecraft Modes

The definition of spacecraft modes is an important aspect in the satellite design process and need to be documented in detail in the satellite user manuals being the main source of information during operations. The ground segment systems engineer needs to also have a thorough understanding of all available satellite modes, subsystem modes, and possible transitions in order to ensure a fully compatible GCS architecture. The configuration of the system modes needs to be

---

[7] Heaters are more frequently deployed inside propulsion subsystems (e.g., fuel lines, valves, etc.) or for the depassivation of batteries.

tailored to the specific needs and circumstances of the satellite during the following main operational phases:

- The pre-launch phase with the satellite located in the test centre and connected to the Electronic Ground Support Equipment (EGSE) for testing purposes.
- The launch and early orbit phase (LEOP) comprising launch, ascent, S/C separation from the launch vehicle, activation of the satellite transponder, deployment of solar arrays, and the satellite's placement into the operational orbit with the correct attitude.
- The commissioning phase during which the in-flight validation of all subsystems is performed.
- The nominal operations and service provision phase.
- The end-of-life phase during which the satellite is placed into a disposal orbit and the satellite is taken out of service.

Following the phases mentioned above, the following satellite system modes would be adequate:[8] during the launch, the satellite is put into a *standby mode*, in which the OBC is booted and records a basic set of house-keeping telemetry. The power generation must be disabled (solar array stowed and battery not connected) and any AOCS related actuators deactivated. Once ejected from the dispenser, the satellite enters into the *separation mode*, in which it stabilises its attitude, reduces rotational rates, starts transmission to ground, and deploys its solar arrays oriented to the Sun to initiate the power generation. Depending on the mission profile and its operational orbit, a set of orbit correction manoeuvres are usually required during LEOP to reach the operational target orbit. Such manoeuvres are also required during the nominal service phase and referred to as station keeping manoeuvres. The execution of manoeuvres usually requires a reorientation of the satellite and also implies additional requirements on the attitude control system to guarantee stability. Therefore, a dedicated system mode for the execution of manoeuvres is usually foreseen which is referred to as *orbit correction mode* or OCM. The *nominal mode* is used for standard operations and is optimised to support the payload operations that provide the service to the end user (e.g., pointing of the high gain antenna to nadir direction).

A very important mode is the *safe mode* (SM) which is used by the satellite in case a major failure is detected. The SM is designed to transition the satellite into a state in which any further damage can be avoided. This comprises a safe orientation, a reduction of power consumption, and the transmission of critical housekeeping telemetry to the ground segment. Potential reasons for such a transition into SM could be subsystem failures, an erroneous TC from ground, or, in a worst case scenario, even a complete black-out of the entire ground segment implying a loss of ground contact. The transition to SM can either be commanded from ground or initiated autonomously by the OBC through its *Failure Detection Isolation and*

---

[8] The system mode naming convention presented here is quite generic and will most likely differ from one satellite to the other. The satellite specific user manual should therefore be consulted.

Recovery (FDIR) program. The capability of an autonomous SM transition is a very important satellite functionality, as the time needed to realise a satellite malfunction on ground and the time to take corrective actions will usually be too long to avoid further damages to the satellite. Once in SM, the satellite can survive without ground interaction for a certain amount of time. It however needs to always stay reachable and available to receive telecommands allowing to recover it and initiate a transition back into nominal mode.

3.3   The Satellite Life Cycle

The development of a satellite follows a typical life cycle which is schematically depicted in Fig. 3.10. It is important for the ground segment systems engineer to be very familiar with these phases and related milestones in order to be able to align and coordinate the ground segment development schedule. Similar to every technical project the first phases of the satellite life cycle deal with the consolidation of requirements at project level involving the customer and relevant stakeholders to ensure a thorough understanding of all end user needs and expectations and to capture these into a concise, accurate, and complete set of project or system requirements. These need to be documented in the *project requirements document* (cf. [16]) which serves as the top level source of information applicable to all stakeholders in the project. The project requirements are used as a basis to derive

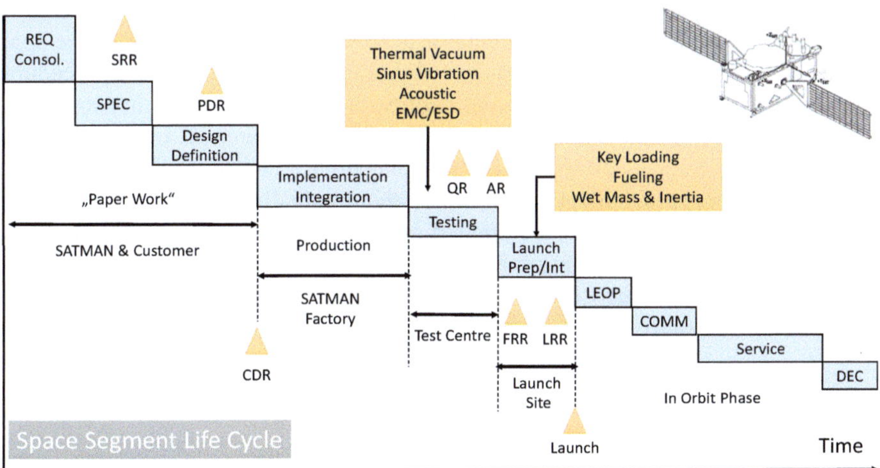

Fig. 3.10 Overview of satellite life cycle and relevant milestones. REQ = Requirements, SRR = System Requirements Review, PDR = Preliminary Design Review, QR = Qualification Review, FRR = Flight Readiness Review, LRR = Launch Readiness Review, EMC = Electromagnetic Compatibility, ESD = Electrostatic Discharge, LEOP = Lauch and Early Orbit Phase, COMM = Commissioning Phase, DEC = Decommisioning Phase

lower level requirements which are technically tailored to each segment and their respective subsystems. The lower level requirements can then be used as a *specification* for the satellite platform and its payload which is an important input for the design phase. The term *consolidation* implies several steps of iteration in order to converge to an acceptable version which can be formalised and "frozen" at the *system requirements review* (SRR) milestone at which the formal *requirements baseline* is established. This means that any subsequent change of a requirement needs to be subject to an agreed configuration control process[9] in order to properly assess the technical, schedule, and cost impact.

In the following phase the satellite architecture is defined and documented and thoroughly scrutinised at the *preliminary design review* (PDR) and the *critical design review* (CDR) prior to the production of real hardware. During the assembly phase the satellite will usually be located at the satellite manufacturer's premises. Some of the subsystems might be subcontracted to other companies which will then deliver them to the prime contractor for integration.

The subsequent phase is the testing phase which is performed at subsystem and system level. Typical system level test campaigns are the thermal-vacuum, vibration, acoustic, and EMC tests which require the relocation of the satellite to a specialised test centre facility that is equipped with the necessary infrastructure like vacuum chambers, shakers, or noise cancellation rooms. Due the high cost to build, maintain, and operate such a complex test facility, they are usually not owned by the satellite manufacturer but rented for specific test campaigns.

A very important test is the *system compatibility test* during which the satellite's OBC is connected to the *Electronic Ground Support Equipment* (EGSE) which itself is connected to the mission control system (SCF) of the ground segment. This setup allows the ground segment engineers and operators to exchange representative TM and TC with the OBC and to demonstrate the full compatibility of the space and ground segments together with the validation of the flight control procedures. Environmental testing is performed at *qualification* level which demonstrates that the satellite performs satisfactorily in the intended environment even if a *qualification margin* is added on top of the nominal flight conditions. This is followed by testing at *acceptance* level during which the nominally expected environmental conditions (with no margins) are applied aiming to demonstrate the conformance of the product to its specification and to discover potential manufacturing defects and workmanship errors (cf. [17]).[10] At the end of all the test campaigns a *qualification review* (QR) is held to analyse and assess the results and if no major

---

[9] A configuration control process usually involves the issuing of configuration change requests, their evaluation by an engineering board, and a final approval by an established Configuration Control Board or CCB.

[10] Qualification test campaigns are usually performed on a dedicated *qualification model* (QM) that is produced from the same drawings and materials as the actual *flight module* (FM), whereas acceptance testing is done with the actual FM.

non-conformance is discovered the satellite can pass the *acceptance review* (AR) which is a prerequisite for its transportation to the launch site.

At the launch site the flight keys are loaded, the tanks are fuelled, and final measurements of wet mass, sensor alignments, and inertia matrix are performed. After the launch site functional tests have confirmed that the satellite performance has not been affected by transportation, the *flight readiness review* (FRR) can be considered successful and its integration into the launcher payload bay can be started. An important aspect of this process is the verification of all launch vehicle payload interface requirements, ensuring that the satellite is correctly "mated" with the launcher both from a mechanical and electrical point of view. The final review on ground is the *launch readiness review* (LRR) which gives the final "go for flight".

Once the satellite is ejected from the launcher's upper stage, the LEOP phase starts during which the satellite needs to establish first contact with ground, acquire a stable attitude, and deploy its solar panels. The LEOP usually comprises a set of manoeuvres to reach the operational orbit. Once the satellite has acquired its nominal attitude, a detailed check up of all subsystem and the payload is performed. This phase is referred to as satellite commissioning and proofs that all satellite subsystems have survived the harsh launch environment and are fully functional to support the operational service phase.

Once the satellite has reached its end of life (usually several years), the decommissioning phase starts. Depending on the orbit height, a de-orbiting or manoeuvring into a disposal orbit is performed.

3.4 Ground to Space Interface

The ground-to-space interface definition is one of the most important activities in every space project as it defines all the parameters which ultimately enable the communication link between the satellite and the ground segment and are therefore relevant for both the satellite and the ground segment systems engineer. For the satellite, the ground-to-space ICD will mainly impact the transponder design (e.g., RF front end) and the OBC (TM/TC protocol) and at ground segment level the TT&C station subsystems (e.g., RF chain and BBM) and the SCF (TM/TC protocol) will be affected. Examples of typical parameters defined in such an ICD are summarised in Table 3.1.

Another important definition is the *physical layer operations procedure* (PLOP) which represents a protocol that defines the necessary steps for the TT&C station to establish the physical layer contact to the satellite. A PLOP is usually characterised by a specific time series of *carrier modulation modes* (CMM) as shown in Fig. 3.11 and has been standardised by the Consultative Committee for Space Data Systems (CCSDS) in their dedicated standard CCSDS-201-B-3 (cf. [18]).

Table 3.1 Examples of parameters defined in the Ground to Space ICD

Parameter	Examples
Modulation types	PM, BPSK, Spread spectrum (DSS), etc.
Ranging types	Code ranging, tone ranging
Carrier specification	Frequency (uplink/downlink), modulation index, linearity requirement
Subcarrier	Frequency, symbol rate (nr. symbols per sec), frequency stability
Baseband characteristics	Signal waveform (e.g., NRZ-L), transition density, data bit jitter, symbol rise time, BER
TT&C RF-performance	Minimum EIRP, Carrier frequency resolution, frequency stability, receive G/T, polarisation, axial ratio, Doppler compensation, maximum supported Doppler shift, Phase noise, Gain slope, Carrier acquisition time, etc.
TT&C Tracking	Maximum tracking and pointing errors, tracking modes
S/C RF-performance	S/C EIRP and stability, polarisation, minimum receive G/T, turn-around ratio, gain, gain slope, gain flatness, etc.
S/C Transponder modes	Coherent & non-coherent modes

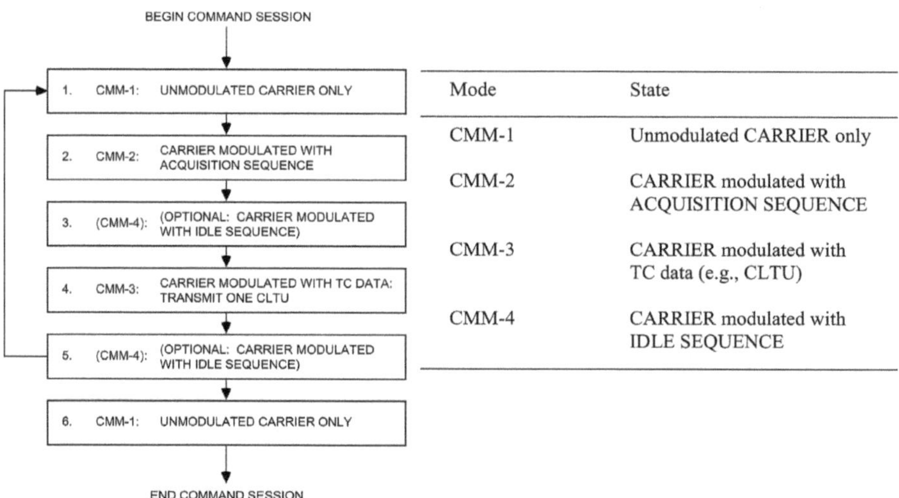

Fig. 3.11 Physical Layer Operations Procedure (PLOP-1) as outlined in Fig.4-1 and Table 4-1 of CCSDS-201.0-B-3 [18]. Reprinted with permission of the Consultative Committee for Space Data Systems © CCSDS

References

1. Sforza, P. (2012). *Theory of aerospace propulsion* (Chap. 12, pp. 483–516). Amsterdam: Elsevier Science.
2. Fortescue, P., Stark, J., & Swinerd, G. (2003). *Spacecraft systems engineering*. Hoboken: Wiley.
3. Hayakawa, H. (2014). Bepi-Colombo mission to Mercury. 40th COSPAR Scientific Assembly.

4. Wertz, J. R. (2002). *Spacecraft attitude determination and control.* Dordrecht: Kluwer Academic Publishers.
5. Sidi, M. J. (1997). *Spacecraft dynamics and control, a practical engineering approach.* Cambridge: Cambridge University Press.
6. National Aaeronatics and Space Administration. (1969). Spacecraft Earth Horizon Sensors. SP-8033.
7. Chu, Q. P., van Woerkon P. T. L. M. (1997). GPS for low-cost attitude determiniation, a review of concepts, in-flight experiences, and current developments. *Acta Astronautica, 41,* 421–433.
8. Eickhoff, J. (2012). *Onboard computers, onboard software and satellite operations. Springer series in aerospace technology.* Berlin: Springer. ISSN 1869-1730.
9. US Department of Defense. (1987). Digital time division command/response multiplex data bus. MIL-STD-1553B.
10. European Cooperation for Space Standardization. (2019). SpaceWire -Links, nodes, routers and networks. ECSS-E-ST-50-12C, Rev.1.
11. International Organization for Standardization. (2003). International standard, road vehicles - controller area network (CAN) – Part 1: Data link layer and physical signalling. ISO 11898-1.
12. European Space Agency. (1994). Data for the selection of space materials. PSS-01-701, Issue 1, Rev.3.
13. European Space Agency. (1989). Spacecraft thermal control design data. PSS-03-108, Issue 1.
14. Redor, J. F. (1990). Introduction to spacecraft thermal control. ESA EWP1599, Version 1.10.
15. Peterson, G. P. (1994). *An introduction to heat pipes - Modelling, testing and applications.* Hoboken: Wiley.
16. European Cooperation for Space Standardization. (2009). Space project management, Project planning and implementation. ECSS-M-ST-10C Rev.1.
17. European Cooperation for Space Standardization. (2002). Space engineering, testing. ECSS-E-10-03A.
18. The Consultative Committee for Space Data Systems. (2000). Telecommand channel service Part 1. CCSDS 201.0-B-3, Blue Book.

Chapter 4
The Ground Segment

The detailed design of a ground control segment (GCS) will most likely differ throughout the various space projects but one can usually find a common set of subsystems or elements responsible to fulfil a clear set of functional requirements. A quite generic and simplified architecture is depicted in Fig. 4.1 showing elements that address a basic set of functions described in Sect. 4.1. These functions are realised in specific elements indicated by the little yellow box with the three letter acronyms (see also Figure caption) which are then described in more detail in the following Chaps. 5–10. Rather generic element names have been used here and the actual terminology will differ in the specific GCS realisation of a space project. It could also be the case that several elements or functionalities are combined into one single application software, server, or rack and it is therefore more adequate to think of each described element here as a component that fulfils a certain set of tasks. The actual realisation could either be a specific software application deployed on a dedicated server or workstation or a virtual machine and will in the end depend on the actual segment design and its physical realisation.

4.1 Functional Overview

4.1.1 Telemetry, Tracking and Commanding

The *telemetry tracking and commanding* (TT&C) function provides the direct link between the satellite and the ground segment. In most space projects the elements implementing this function will be located at a geographically remote location from the remaining GCS infrastructure. It has to support the acquisition of telemetry (TM) from the satellite and to forward it to the GCS. In the opposite direction, it receives telecommands (TC) from the GCS and transmits them to the satellite. In addition it has to support the generation of radiometric data which comprise the measurement

© The Author(s), under exclusive license to Springer Nature Switzerland AG 2023

B. Nejad, *Introduction to Satellite Ground Segment Systems Engineering*, Space Technology Library 41, https://doi.org/10.1007/978-3-031-15900-8_4

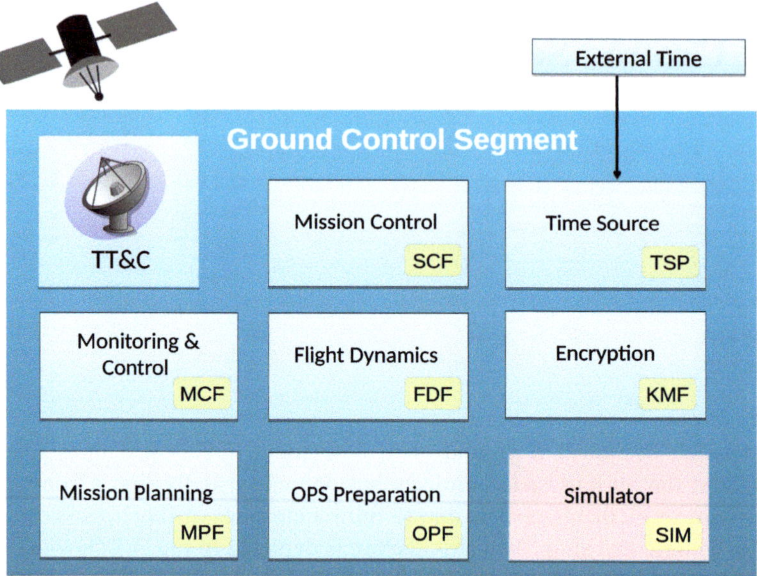

Fig. 4.1 Ground control segment (GCS) functional architecture. The boxed acronyms refer to the element realisation as described in Chaps. 5–10, i.e., FDF = Flight Dynamics Facility, KMF = Key Management Facility, MCF = Mission Control Facility, MPF = Mission Planning Facility, OPF Operations Preparation Facility, SCF = Satellite Control Facility, TSP = Time Service Provider

of the distance between the satellite and the antenna phase centre (ranging), the angular position of the satellite location projected onto the celestial sphere (angles), the relative velocity between the satellite and the TT&C station (Doppler), and finally some measurement of meteorological data. The latter one typically comprise temperature, pressure, and humidity (at the location of the ground station) which are needed for the modelling of the signal delay time in the Earth's atmosphere. Whereas the acquisition of TM and transmission of TCs are linked to the Satellite Control Facility (SCF), the radiometric data are needed by the Flight Dynamic Facility (FDF) as an essential input for the determination of the satellite and orbit geometry and its position.

The number of TT&C stations needed for a space project depends mainly on the mission goals. If more than one ground station is being involved the term *ground station network* is used. The amount of time a satellite is visible for a specific TT&C station is highly dependent on the orbit altitude, its shape,[1] and its inclination with respect to the Earth equator. Whereas satellites visibilies in Low Earth Orbit (LEO) last up to a few minutes a satellite in Medium Earth Orbit (MEO) or in Geosynchronous Orbit (GEO) can last up to several hours.

---

[1] The shape of the orbit is expressed by the so called orbit eccentricity which can range from close to circular to highly elliptic.

The ground-track of the satellite will strongly depend on the orbit inclination.[2] This orbit element also determines the highest latitude a satellite can be seen from ground which puts a constraint on the usability of a certain ground station for a satellite contact.

The need for ground station coverage, i.e., the required amount of time a satellite is visible from ground, varies throughout the operational phase the satellite is in. During LEOP the execution of orbit manoeuvres could require double station visibility for contingency planning reasons, whereas during routine operations only one contact per orbit might be sufficient. The required contact duration for routine contacts itself is often driven by the specific payload needs to downlink (dump) its telemetry and the available data rate provided by the space to ground link design.

From a flight dynamics point of view, ground station coverage (or the lack thereof) has an impact on the achievable orbit determination accuracy. For this both the geographical distribution of the ground stations around the globe providing visibility of several different arcs of an orbit and the duration of a satellite contact determining the length of the orbit arc have an influence (see Chap. 6).

4.1.2 Flight Dynamics

The main responsibility of the Flight Dynamics function is to maintain an up-to-date knowledge of all satellite orbits and from this derive all orbit dependent products that are required for daily operations.

To enable a TT&C antenna to establish a satellite contact, it requires the so called *pointing information* allowing it to direct its feed to the correct location on the sky where the satellite is located. Flight dynamics therefore has to generate pointing files which are based on the most recent orbit knowledge and are typically propagated a few days into the future, providing orbit predictions at the time of contact.

Depending on the service offered, most satellite projects require the satellites to reach a predefined operational target orbit during their LEOP phase and to retain a predefined orbit position or slot with a defined accuracy throughout its lifetime. This required orbit position could either be defined as a stationary control box like in the case of a geostationary satellite or as moving orbital slots in the case of satellite constellations.

Satellites flying in LEO are usually equipped with some type of remote sensing payload which require orbits that either fulfil a certain constant relative geometry with respect to the Sun (sun-synchronous orbit) or have to keep a specific orbit altitude, or revisit specific ground target sites or regions at defined times (defined by a ground track repeat cycle). In all these cases, flight dynamics has to compute the required orbit correction manoeuvres (Δv) to reach and keep the target orbit and

---

[2] The inclination is defined as the angle between the satellite orbit plane and the equatorial plane of the Earth.

translate these into satellite specific command parameters like thrust orientation, thrust start and end time and/or burn duration.

Many satellite operational activities depend on a number of orbital events like the presence of the Sun or Moon in the field-of-view (FoV) of a sensor, the satellite presence inside or outside the Earth's shadow (eclipse periods), or geometrical conditions like a specific angular collinearity between the Sun, Moon, and the satellite itself. The computation of such events requires orbit knowledge of the satellite but also other celestial planets (mainly the Sun and Moon for Earth orbiting satellite) and are therefore usually computed by flight dynamics.

4.1.3   Mission Control

The main task of the mission control function is to establish the real time communication with the satellite through the reception of telemetry (TM) and the uplink of telecommands (TCs). The telemetry is generated by the onboard computer of the satellite and typically comprises two groups of parameters which stem from two different sources. The first group is referred to as *housekeeping* (HK) telemetry and comprises all the TM parameters that report the status of the various subsystems of the satellite and allow to assess the overall health status of the spacecraft. The second one is the TM collected by the payload and the actual data relevant for the end user of the space asset. It should be kept in mind that the data volume of the payload specific TM can be significantly larger compared to the HK TM. The downlink or *dump* of the payload TM might therefore be scheduled differently to the HK one (e.g., dedicated time slots, contacts, or even ground stations).

Prior to any possible exchange of TM or TC, a communication link between the GCS and the satellite needs to be established. This is usually initiated via the execution of a set of commands that are referred to as *contact directives*. Such directives comprise commands to relocate the TT&C antenna dish to the right location,[3] activate and correctly configure all the required TT&C subsystems in the up- and downlink paths, and finally power-on the amplifiers to transmit (or *raise*) the RF carrier.

The TM is organised in so called TM packets and TM frames that follow an established space communication protocol. To avoid the need for every satellite project to develop its own protocol, the need for protocol standardisation has been identified already in early times. One of the most widely used space communication protocols is the *Space Link Extension* or SLE protocol which has been standardised by the Consultative Committee for Space Data Systems (CCSDS) [1] since the 1980s and is explained in more detail in Sect. 7.2.

---

[3] The correct satellite location and track on the sky for a given TT&C station needs to be pre-computed by the FDF and provided to the station beforehand.

Mission control is responsible for the transmission of all TCs to the satellite. The purpose and complexity of TCs cover a very broad range that can be a simple test TC verifying the satellite link (usually referred to as *ping* TC) up to very complex ones that comprise hundreds of different TM parameters (e.g., a navigation message of a GNSS satellite). Most of the tasks that need to be carried out require the transmission of a set of TCs that need to be executed by the satellite in the right sequence and timing. This is usually achieved through the use of predefined TC *sequences* which are loaded into the TC *stack* in the required order before being released (uplinked) to the spacecraft. This allows the *satellite operations engineer* (SOE) to perform the necessary cross-checks prior to transmission.

After reception on ground, all new TM needs to be analysed which is one of the main responsibilities of the SOE. As there are usually hundreds of different TM parameter values to be scrutinised after each contact, the SCF provides a computerised cross check functionality which compares each received TM parameter to its expected (i.e., nominal) value and range. The definition of the expected range requires a detailed understanding of the satellite architecture and its subsystems and is therefore an important input to be provided by the satellite manufacturer as part of the "out of limit" database. This database needs to be locally stored in the SCF so it can be easily accessed. In case an out-of-limit parameter is detected, it needs to be immediately flagged to the operator in order to perform a more detailed analysis.

Even if not mandatory for operations, the mission control function should provide means for an in-depth analysis of the satellite TM received on ground. This capability should comprise tools for the investigation of trends in time-series as well as cross correlations between different types of TMs. The ability to generate graphical representations will further support this. Also the ability to compute *derived* TM parameters, defined as parameters that are computed from a combination of existing TM values, can be helpful to gain a better understanding of anomalies or events observed on the satellite.

Another key functionality is the ability to support the maintenance of the satellite onboard software (OBSW) in orbit which is an integral part of the satellite and therefore developed, tested, and loaded into the satellite's onboard computer (OBC) prior to launch. In case a software problem is detected post launch, the OBSW will require a remote update and the satellite manufacturer has to develop a patch or even a full software image and provide it to the ground segment for uplink. The ground segment needs to be able to convert this patch or image into a TC sequence that follows the same packet and frame structure as any routine TC sequence used during nominal satellite operations. A verification mechanism at the end of the transmission should confirm that the full patch or image has been received without errors and could be stored in the satellite onboard computer.

More and more space projects today rely on satellite constellations which puts a high demand on the daily operational workload. Especially for satellites that have reached their operational routine phase, the TM/TC exchange will follow a very similar pattern that does not require a lot of human interaction. Automated satellite monitoring and control has therefore become a key functionality to reduce the workload and operational cost in a satellite project. An essential input for the automation component is the reception of a schedule that contains all the upcoming contact times and expected tasks to be executed within each of the contacts.

4.1.4   Mission Planning

Successful satellite operations is highly dependent on the efficient management of resources considering applicable constraints and the ability to coordinate activities and resolve conflicts. The main goal of every mission planning function must be to optimise the service time of a satellite and its payload in order to maximise the benefit for the end user or costumer. This goal however needs to be consolidated with a number of limitations and resource constraints. Typical resource limitations are limited satellite contact times (mainly dependent on the orbit geometry), the number of available ground stations and their geographical distribution, the capacity of the onboard computer data storage, the downlink data rate of the satellite transponder, the available fuel budget (determining the satellite lifetime), and any satellite limitations affecting its daily operation which are usually expressed as *flight rules*. A typical example for the need to coordinate conflicting activities are requests for observation time on satellites with space telescopes. An observation request from a customer might not be possible if it is in conflict with an ongoing or planned maintenance activity which limits the use of either the satellite or the ground segment during that time.

The coordination of all satellite and ground activities is the domain of *mission planning* and the main inputs and outputs of a generic planning process is graphically depicted in Fig. 4.2. The most important result of that function is a mission timeline or plan or a (computer readable) contact schedule that can be directly processed by the SCF for automated operations. Whatever format, the resulting plan needs to fulfil the following requirements:

- all *planning requests* (PRs) were considered and any conflict could be resolved according to the known priorities and constraints;
- the plan respects all known constraints and resource limitations;
- the plan optimises the use of the satellite and its payload.

An important input to the planning process are the times of ground station visibility to a satellite and any relevant orbit events. Examples for such orbit events are Earth and Moon eclipse times and duration, ascending and descending node crossings, satellite collinearities with the Sun and the Moon, or sensor blinding events. As the computation of this information requires the most recent knowledge of the satellite state vector (orbit position and velocity), orbit events are usually computed by flight dynamics and provided to mission planning as a so called *event file*.

Fig. 4.2 The mission planning process, overview of inputs and outputs

A PR can come from either the mission planning operator, another element inside the GCS (e.g., updates of the onboard orbit propagator, or manoeuvre planning inputs), or even a project external source or community. In scientific satellite projects (e.g., space telescopes) such external request are typically requests for observation time that specify the type, location, and duration of an observation and can also imply operational tasks for the ground segment (e.g., change of the satellite attitude to achieve the correct pointing or switch to a special payload mode). To ensure that all the relevant information is provided and to ease a computerised processing of this input a predefined format or template for a PR should be used.

As already mentioned, any planning activity has to consider all applicable resource constraints and contact rules which is another important input to the planning process. Some might be more stringent than others which needs to be clearly identified. Typical examples for planning constraints are:

- maximum and minimum contact duration,
- maximum and minimum time spans between consecutive contacts,
- available downlink data rate (determining the required time to downlink a given data volume),
- mission important orbit location (e.g., for Earth observation purposes),
- any payload and platform constraints which have an impact on the satellite operation.

4.1.5 Operations Preparation

Reliable satellite operations is strongly depend on the correct ground segment configuration and the correct development, validation, and storage of all operational products like the satellite reference databases or flight and ground segment operations procedures. The ground segment configuration and the management of operational products is best achieved via the definition of a set of configuration files that are distributed among all the ground segment components or elements. As any uncontrolled and incorrect modification of this sensitive data items could have a negative impact on the operational behaviour or the correct generation of a component's products and potentially even jeopardise the overall GCS operability, the implementation of a configuration control process and tool inside the GCS is essential. This requires the categorisation of all configuration data into so called *configuration items* or CIs based on content, data source (i.e., who has generated this data set) and end user (i.e., which element uses that data). The clear identification of the source (or provider) of a CI is very relevant for those that are provided by an external entity like the satellite manufacturer (e.g., satellite specific data on mass, fuel, thrust, inertia usually used by flight dynamics).

4.1.6 Monitoring and Control

In order to guarantee uninterrupted satellite operations, the functional status of the entire ground segment needs to be continuously measured and reported. Any malfunctioning of an operationally critical function or element has to be detected early enough for an operator to take corrective action and to prohibit any damage to the spacecraft, its payload, or interruption of service to the end user. This monitoring needs to be done at both software and hardware level. Examples of typical hardware parameters highly relevant for the proper functioning of the element are disk space, memory or CPU usage, or rack temperatures.

Another important feature is the *event log* which represents a time sequential view of all events that have occurred. The criticality of events is usually indicated by a specific colour code (e.g., green, yellow, red) and for very critical ones could even be accompanied by an audible tone or message to stress that an immediate operator response is needed.

In addition to the collection of all the relevant monitoring data according to the defined interfaces, its presentation to the operator and the ability to archive all the information for later investigation needs to be considered. Due to the huge amount of data that needs to be scrutinised by the ground segment operators on a daily basis, the efficient and ergonomic design of the Man Machine Interface (MMI) layout deserves special attention and will therefore be addressed in more detail in the element specific chapter (refer to Chap. 10).

4.1.7 Simulation

The development and use of simulators is essential throughout all phases of a project, starting from early feasibility studies, followed by requirements specification, verification and validation, and finally during operations. The terminology used for the simulator often depends on the project phase and its exact purpose. The European Cooperation for Space Standardization (ECSS) provides a dedicated standard referred to as ECSS-E-TM-10-21A [2] which describes all kind of simulators used in space projects and also suggests a terminology for their clear distinction. In the context of ground segment systems engineering the following two simulator types are relevant:[4]

- The *ground system test simulator* or briefly *AIV simulator* is needed during the segment integration, verification, and validation phases when special attention needs to be given to achieve the full compatibility between the ground and space segments.
- The *training, operations and maintenance simulator* or briefly *operational simulator* is required for the proper validation of all operational procedures which are developed by the flight control team (FCT) before they can be applied in actual satellite operations.

In both cases the simulator has to offer a near real time behavioural model of the satellite platform and its payload. The development and deployment of one single simulator which addresses the requirements of both, the AIV and operational simulator, is highly beneficial as it will avoid any inconsistent behaviour when the same equipment is used by the ground segment developers and operators and furthermore reduces development time and cost. Following this approach, there is no distinction made between the AIV and operational simulator presented in Chap. 11 where the term *satellite simulator* or SIM is simply used to refer to both types equally.

No matter how sophisticated a simulator design is, it is important to keep in mind that there will always remain a certain gap concerning its representativeness to the real satellite. It is therefore important that the validation of operational products is backed up by test campaigns that involve the satellite Flight Module (FM) as long as it is still accessible in the test centre of the satellite manufacturer.[5]

---

[4] The set of simulators used during the early project definition phases (e.g., system concept simulator, functional engineering simulator, or functional validation testbench) and the ones developed by the space segment (e.g., spacecraft AIV simulator or EGSE, onboard software validation facility) are not described here and the reader is referred to ECSS-E-TM-10-21A [2].

[5] The connection from the GCS to the satellite onboard computer would then be achieved via dedicated Ground Support Equipment (EGSE).

4.1.8   Encryption

In some projects the satellite might be equipped with a payload or platform module that requires encrypted commanding and provides security sensitive data to the end user. In this case the ground segment needs to host an encryption function able to perform the encryption operations through the implementing of the specific algorithms and a set of keys that need to be known to both the ground and the space segment. At satellite level this is ensured through a dedicated key loading operation which is usually performed a few days before the satellite is being lifted and mounted into the launch vehicle payload bay. To ensure that all encryption operations can be performed efficiently with minimum latency being introduced, a close interaction with via a secure interface to the mission control function needs to be realised. Due to the very project specific design and realisation and its classified nature no detailed chapter is provided and the reader is recommended to consult the project specific design documentation.

4.1.9   Timing Source

All ground segment functions need to be synchronised to the same time system and one single time source. This can either be generated inside the ground segment or provided by an external source referred to as the *time service provider* (TSP). For navigation (GNSS) type projects such an external source is not required as an accurate time source is usually available from the atomic clock deployed in the ground segment which is also the basis for the definition of the time system (e.g., GPS or GST time). Other satellite projects require an external time system which needs to be distributed or synchronised to all the servers that host the various functions. A convenient way to distribute timing information is to use the network time protocol (NTP) that can take advantage of the existing network infrastructure that is already deployed. NTP is a well established time synchronisation protocol used on the internet and is therefore supported by the majority of operating systems in use today.

4.2   Physical Architecture

It is important to distinguish between the *functional* and the *physical* architecture of the ground segment. The functional architecture (see Sect. 4.1) describes the specific tasks each of the GCS components needs to perform, or in other words, which functional requirements they are supposed to satisfy. The physical architecture describes the actual realisation in terms of hardware and software deployment which comprises the racks and servers mounted inside, the rack deployment in the server rooms, the number of workstations and their distribution in the control rooms, the network design, storage, and virtualisation concepts (see Chap. 13).

4.3 Interface Definition

The exchange of information inside the GCS but also with external entities can only be realised through the proper definition and implementation of interfaces. An accurate and complete interface specification requires as a minimum the definition of the following points

- the exact scope of information that needs to be exchanged,
- the format that needs to be used for the exchanged information (contents and filename),
- the exchange mechanism and protocol to be used,
- the specification of all relevant performance parameters (e.g., file size, frequency of exchange).

The definition of a file content can comprise both required and optional information. Optional input must be clearly identified as such and a recommendation given on how to use it in case provided. The file format can be binary or ASCII, where the latter one has the obvious advantage to be human readable. In modern ground segment design the *Extensible Markup Language* (XML) has gained significant importance and is extensively used in the design of interfaces. XML brings two main advantages, a clearly defined syntax (file format) and an extensive set of rules that allows to specify the expected file content and constraints. Examples of the latter are the expected data format (e.g. string or numerical value), the allowed range of a value, or limitations on the number of data items to be provided in a file. As these rules are defined as part of the XML schema, they can be easily put under configuration control which is essential for the definition of an interface baseline. The exchange mechanism and protocol should clarify how the file transfer must be done. Whereas in some cases a simple file transfer protocol (ftp) might be adequate, it might not fulfil the security needs in other cases.[6]

The interface performance specifies the data volume that needs to be transferred over the network which is determined by the average file size of each transfer and the transfer frequency. An accurate assessment of the required interface performance during operations is an important input for the correct sizing of the network bandwidth. An underestimated bandwidth can cause a slow down in data transfer that could impact the overall ground segment performance.

Every ground segment architecture will require the definition and implementation of a number of interfaces both internally but also to external entities which is schematically shown in Fig. 4.3. Files and signals that need to be exchanged with the same entity can be grouped and specified in one interface control document indicated by ICD-A, ICD-B, or ICD-C. The term *control* emphasises the need for configuration control in addition to the interface specification. To guarantee a flawless flow of information, the same ICD versions need to be implemented on both

---

[6] More elaborate file transfer mechanism are presented in Sect. 12.2.

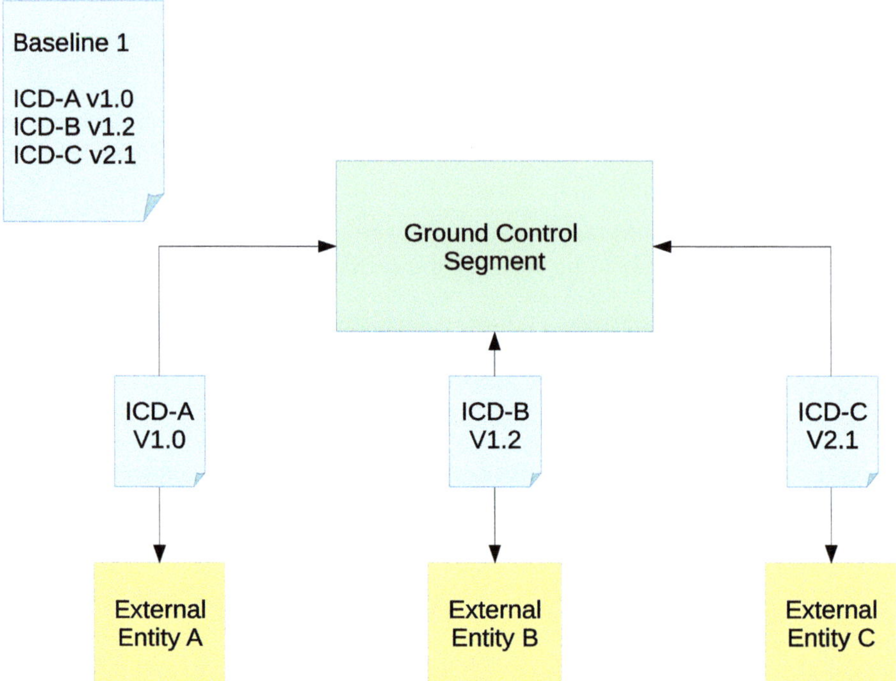

Fig. 4.3 Example of interface specification with three external entities A, B, and C. Each interface control document (ICD) is put under configuration control via a specific version number. The interface baseline ("Baseline-1") is defined via a set of ICDs and their respective version numbers

sides of an interface. Even if this seems obvious, it might not always be a simple task if the two ends are under different contractual responsibilities and implemented by different entities. The definition of an interface baseline (e.g., "Baseline 1" in Fig. 4.3) specifies a set of documents and their expected version numbers and is a useful means to make them applicable to a subcontractor. This can be formally done via the *contractual item status list* or CISL which is issued by the contracting entity and the contracted entity can confirm the correct reception and implementation by issuing the *contractual status accounting report* or CSAR.

References

1. The Consultative Committee for Space Data Systems. (2007). Overview of space communication protocols. CCSDS 130.0-G-2, Green Book.
2. European Cooperation for Space Standardization. (2010). Space engineering, system modelling and simulation. ECSS-E-TM-10-21A.

Chapter 5
The TT&C Network

The TT&C station's main functions can be derived from its acronym which refers to *telemetry* (transmission), *tracking*, and *control* (see Fig. 5.1). The word telemetry reception should be seen in a wider context here comprising the satellite's own housekeeping data that provide the necessary parameters to monitor its health and operational state but also the data generated by the payload. The control part refers to the ability of the TT&C station to transmit telecommands (TCs) to the satellite allowing the GCS to actually operate the satellite.

The overall TT&C station architecture can be decomposed into three main building blocks, (1) the mechanical part providing the ability to support the heavy antenna dish and point it to a moving target in the sky, (2) the RF subsystem comprising all the subsystems of the uplink and downlink signal paths, and (3) the supporting infrastructure comprising the station computers, network equipment, and power supplies in order to monitor and operate the station from remote and provide a stable link for the transfer of all data and tracking measurements to the control centre.

5.1 Tracking Measurements

Satellite tracking is a fundamental function of every TT&C station as it provides the measurements needed as input to the orbit determination function providing a regular update to the satellite's current orbit knowledge (see also Chap. 6). The following types of tracking data are usually generated by a TT&C station:

- *Slant range* measurements ρ refer to the line-of-sight distance between the ground station and the satellite which are derived from the measurement of the time delay of a pseudo-random ranging code or tone modulated on the uplink-carrier with respect to the one seen on the downlink carrier. This time delay is measured by the station baseband modem using a *delay lock loop* (DLL)

© The Author(s), under exclusive license to Springer Nature Switzerland AG 2023
B. Nejad, *Introduction to Satellite Ground Segment Systems Engineering*, Space
Technology Library 41, https://doi.org/10.1007/978-3-031-15900-8_5

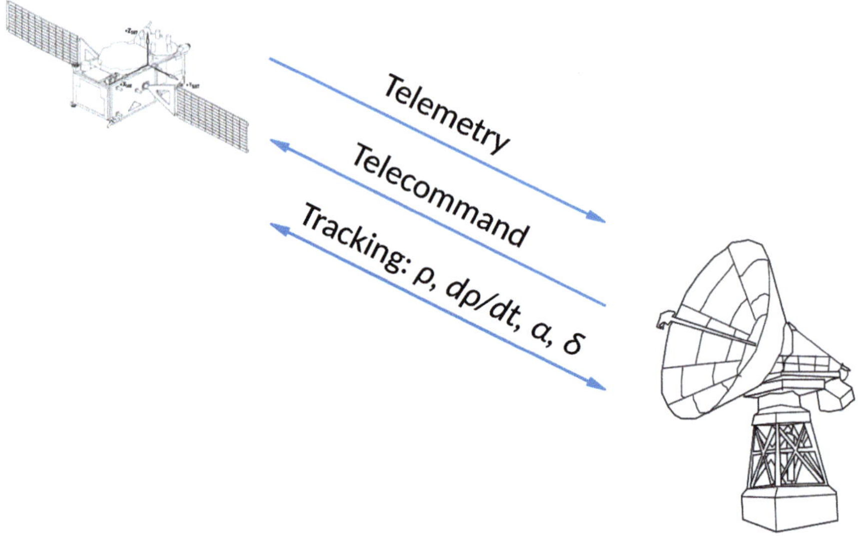

Fig. 5.1 The three main functions of a TT&C antenna expressed by its acronym: tracking, telemetry reception, and command transmission

where the incoming signal is mixed with a locally generated reference signal for which the phase is continuously adjusted (delayed) up to the point of constructive interference. This measured time delay can then be converted into a (two-way) travel distance that provides the range measurement. It is important to keep in mind that the slant range is always measured with respect to phase centre of the satellite and the ground station antennas and additional corrections need to be applied before using them for orbit determination. Such corrections comprise the satellite phase centre to centre-of-mass correction and the subtraction of the additional time delay in the ground station RF equipment (station bias).

- *Angular track* measurements comprise the azimuth and elevation profiles of the satellite orbit projected on the surface of the sky. They are usually measured as function of time during a contact when the antenna is automatically tracking the satellite RF beam. The corresponding operational antenna mode is therefore called *autotrack* mode. The second operational mode is the *program track* mode in which the antenna strictly follows a predicted orbit track which is made available prior to the start of the contact. Every pass usually starts in program track in order to point the antenna to the location where the satellite is expected to appear in the sky. Once the antenna managed to lock onto the satellite signal, it is able to change into autotrack mode which then allows the generation of azimuth and elevation profiles.

- *Doppler* frequency shift measurements Δf are used to determine the relative line-of-sight velocity v_{rel} of the satellite and the station via the simple relationship

$$v_{rel} = \frac{\Delta f}{f_0} c \qquad (5.1)$$

where f_0 is the source frequency (prior to its Doppler frequency shift) and c the speed of light in vacuum. That relative velocity can also be interpreted as the time derivate of the slant range $d\rho/dt$ known as *range rate*. The Doppler shift is measured by the TT&C station receiver by phase-locking onto the incoming signal and comparing the received (downlink) frequency from the satellite to the original uplink frequency. This only works if the satellite transponder works in *coherent* mode which means it multiplies the uplink frequency by a fixed turn-around factor to derive its downlink carrier frequency.[1]

5.2 The Mechanical Structure

The mechanical structure of the antenna assembly can be split into the moving part comprising the large antenna reflector dish and everything mounted to it and the static part consisting of the antenna building that provides the support structure and hosts the motors and bearings that enable the station to move the antenna and meet its pointing requirements. An example of a typical architecture is depicted in Fig. 5.2 showing the generic design of a Galileo TT&C station. The antenna movement needs to be possible in two degrees of freedom referred to as *azimuth* and *elevation*. Azimuth is defined as the motion along the (local) horizon and covers a theoretical span of 360 deg, whereas elevation refers to the movement in the perpendicular direction of 180 deg starting from any point on the horizon crossing the zenith and ending in the exact opposite side on the horizon again. These are of course ideal ranges in a mathematical sense which will not be feasible in reality due to mechanical limitations of the dish and the support infrastructure. To allow a precise positioning at a given azimuth and elevation angle, both the current position and its change needs to be measured via special encoder devices. This information then has to be transmitted to the antenna control unit (ACU) which is the subsystem responsible for the overall management and commanding of the antenna motion.

The antenna motion itself is usually achieved via a set of electrical drive engines that engage in the toothing of the corresponding bearings as presented in Fig. 5.3 showing again the Galileo TT&C ground station design as an example. As the

---

[1] In non-coherent mode the satellite transponder uses its own generated carrier frequency which it derives from its on-board frequency source. This operating mode however does not allow Doppler measurements.

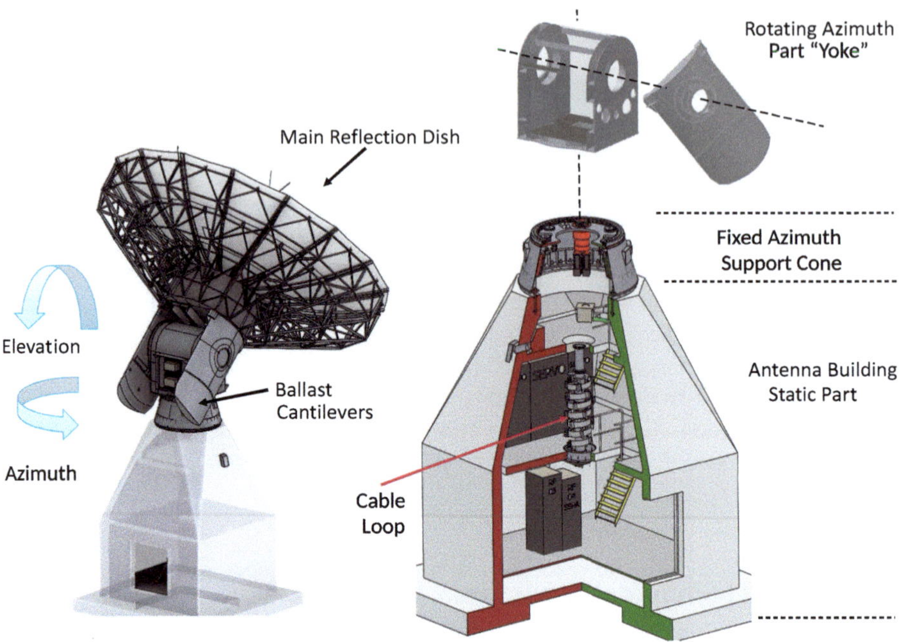

Fig. 5.2 Static and dynamic parts of the Galileo TT&C station. The dynamic part moves in two degrees of freedom, in *azimuth* and *elevation*. The antenna building houses all the RF equipment which is typically mounted into server racks. Also shown is the cable loop for the cable guidance between the static and the moving antenna sections. Reproduced from [1] with permission of CPI Vertex Antennentechnik GmbH

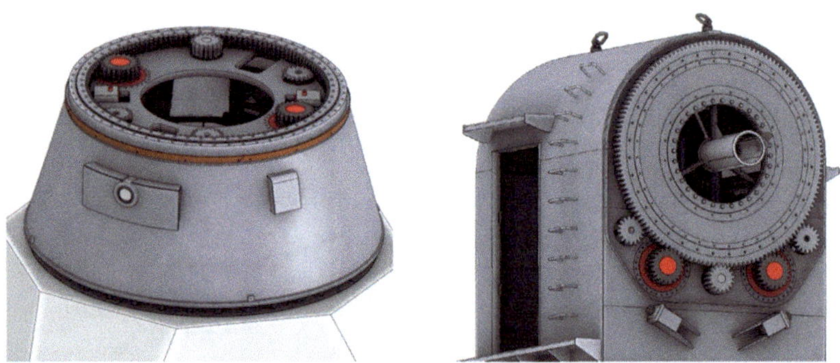

Fig. 5.3 Examples of azimuth and elevation drive engines (coloured in red) from the Galileo TT&C design. Reproduced from [1] with permission of CPI Vertex Antennentechnik GmbH

motion in elevation has to carry the entire weight of the reflector structure, so called *ballast cantilevers* are used as a counter-balance to reduce the mechanical load on the engines during motion.

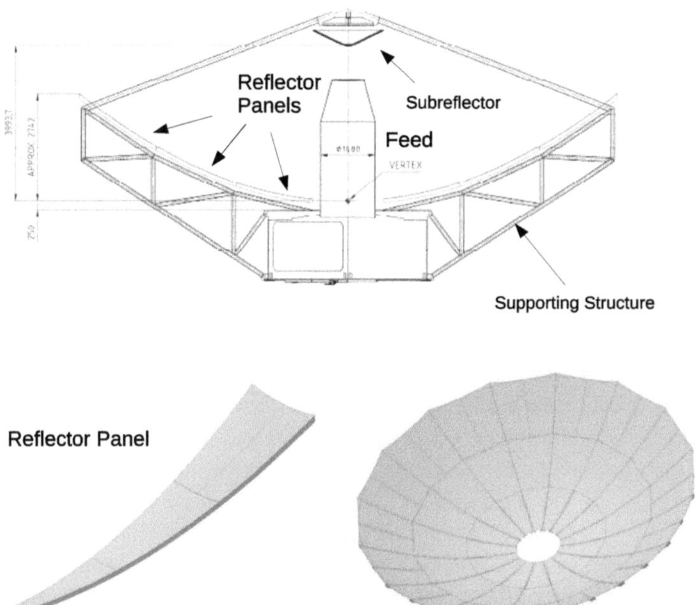

Fig. 5.4 Example of a larger size antenna reflector using an antenna supporting structure and reflector panels. A Cassegrain type dish implements a subreflector that is mounted in the antenna focus and diverts the RF beam into the antenna feed. Reproduced from [1] with permission of CPI Vertex Antennentechnik GmbH

For large antenna diameters (typically above 10 meters) the antenna reflector can be built on top of a supporting structure using individual reflector panels as building blocks as shown in Fig. 5.4. The main advantage is that a large reflector surface can be decomposed into small parts and packed into oversea containers for shipment to a remote location where the dish can then be easily re-assembled. Such a design needs to allow the exact adjustment of each individual panel after it has been mounted onto the support structure in order to achieve an accurate parabolic shape of the full antenna surface. A *Cassegrain* type reflector incorporates a smaller subreflector which is mounted in the focal point of the parabola and reflects the RF beam into the antenna vertex point. From that point onwards a wave guide leads the RF beam into the antenna building where the equipment for the further RF processing is located.

The antenna building structure has to carry the load of the full reflector dish and protect all the sensitive RF equipment from environmental conditions like rain, humidity, storm, and extreme temperatures. The site selection process needs to ensure that the soil for the antenna building foundations ensures sufficient bearing capacity to provide stability during the entire antenna life time. The surrounding soil of the antenna building might even have to provide a higher bearing capacity to support the assembly cranes used during the antenna construction time. In tropical areas close to the equator rain seasons can imply intense loads of water in very short time frames which requires adequate water protection and control to be put in place in order to avoid damage through flooding (e.g., drainage channels). The antenna building also needs to accommodate all the power and signal cables required by the equipment hosted inside. Care must be taken for the cable guidance from the static part to the moving part by using special cable loops as depicted in the centre of the right panel of Fig. 5.2.

5.3   The Radio Frequency Subsystem

The antenna dish shape and its size are a very important design factors as they influence the spatial distribution and power of the radiated energy. The graphical presentation of the radiation properties as a function of space coordinates is referred to as the antenna's *radiation pattern* and its actual characterisation can only be accurately achieved via a dedicated measurement campaign which is relevant to understand both the transmitting and receiving properties of the antenna. Depending on whether the spatial variation of radiated power, electric or magnetic field, or antenna gain is presented, the terms *power pattern*, *amplitude field pattern*, or *gain pattern* are being used respectively.

An ideal antenna that could radiate uniformly in all directions in space is said to have an *isotropic* radiation pattern.[2] In case the radiation pattern is isotropic in only one plane it is referred to as an *omni directional* radiation pattern which is the case for the well known dipole antenna depicted in panel (a) of Fig. 5.5 showing the two perpendicular *principal plane patterns* that contain the electrical and magnetic field vectors $\vec{E}$ and $\vec{H}$. If no such symmetry is present, the antenna pattern is termed to be *directional* which is shown in panel (b) of Fig. 5.5 for the case of a parabolic reflector type antenna.

A directional radiation pattern contains various parts which are referred to as *lobes*. The *main lobe* represents the direction of maximum radiation level and *side lobes* of lower ones. If lobes exist also in the opposite direction to the main lobe they are called *back lobes*. Two important directional pattern characteristics are the *half-power beam width* (HPBW) also referred to as 3-dB beam width and the *beam width between first nulls* (BWFN). The HPBW defines the angular widths within which the radiation intensity decreases by one-half (or 3 dB) of the maximum value whereas the BWFN is defined by a decrease of a factor 10 (10 dB) or even 100 (or 20 dB). The following fundamental properties of the beam width dimension should be kept in mind when choosing an antenna design:

---

[2] Such a uniform radiation pattern is an ideal concept as it cannot be produced by any real antenna device, but this concept is useful for the comparison to realistic directional radiation patterns.

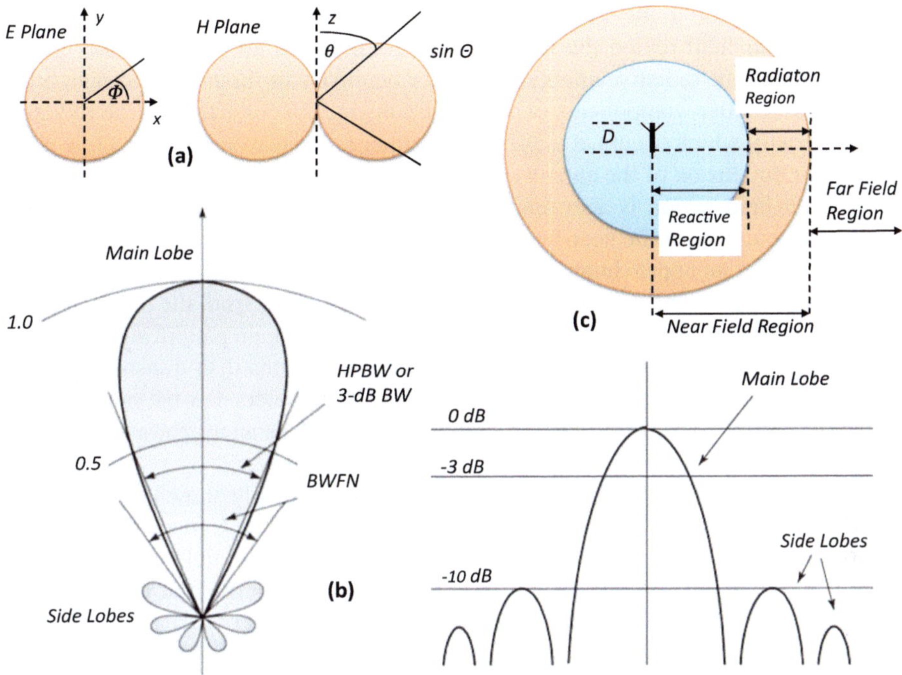

Fig. 5.5 Antenna radiation pattern (far field): (**a**) ideal dipole antenna omnidirectional E- and H-plane radiation pattern polar plot (**b**) directional pattern of parabolic reflector showing main and side lobes, the half-power beamwidth (HPBW), and the beamwidth between first nulls (BWFN); (**c**) definition of antenna field regions with commonly used boundaries

- An increase in beam width will imply a decrease of the side lobes and vice versa, so there is always a trade off between the intended beam width size and the resulting main to side lobe ratio.
- A smaller beam width will provide an improved antenna resolution capability which is defined as its ability to distinguish between two adjacent radiating sources and can be estimated as half of the BWFN.

Another important aspect in antenna theory is the characterisation of the different regions surrounding a transmitting antenna consisting of the *near-field* and the *far-field* regions schematically depicted in panel (c) of Fig. 5.5. The near-field region can be further subdivided into the *reactive near-field* being immediately adjacent to the antenna device and the *radiating near-field*. The first one is characterised by the existence of radiation field components generated by the antenna surface through induction which are referred to as *reactive field components* and add to the radiation field components. Due to the interaction between these two different type of radiation, either the magnetic or the electric field can dominate at one point and the opposite can be the case very close by. This explains why radiation power predictions[3] in the reactive region can be quite unpredictable. In the adjacent radiating near-field region the reactive field components are much weaker giving dominance to the radiative energy. The outer boundary of the near-field region lies where the reactive components become fully negligible compared to the radiation field intensity which occurs at a distance of either a few wavelengths or a few times the major dimension of the antenna, whichever is larger.

The far-field region is also referred to as the *Fraunhofer* region and begins at the outer boundary of the near-field region and extends indefinitely into space. This region starts at an approximate distance of greater than $2D^2/\lambda$ and is characterised by the fact that the angular field distribution is independent from the distance to the antenna. This is the reason why measurements of antenna gain pattern are performed here. To give a simple example, for a 13 meter size antenna dish transmitting in S-Band (ca. 2 GHz), the far-field region starts at ca. 2 kilometers. For the same antenna size transmitting in X-Band (ca. 11 GHz) the equivalent distance grows to more than 12 kilometers.

The description of the antenna field regions above already show the importance of radiation power as an antenna performance parameter. Considering that the energy is distributed equally among the electric and magnetic field components ($\vec{E}$ and $\vec{B}$ with respective units [V/m] and [A/m]), one can derive the surface power density $\vec{S}$ (unit [W/m^2]) building their vectorial cross product

$$\vec{S} = \vec{E} \times \vec{H} \tag{5.2}$$

called *Poynting vector*. The total power P (unit [W]) flowing through a closed surface A is given by the surface integral of the Poynting vector

$$P = \oint\int_A \vec{S} \cdot d\vec{A} \tag{5.3}$$

where $d\vec{A} = \hat{n}\, dA$ is an infinitesimal small surface component with its orientation in space defined by the normal vector $\hat{n}$. Keeping in mind that the electric and magnetic fields are time varying (harmonic) quantities the time-averaged Poynting vector $\vec{S}_{av}$ is the actual quantity of interest.

$$S_{av} = \frac{1}{2} Re\left(\vec{E} \times \vec{H}^*\right) \quad [\text{W/m}^2] \tag{5.4}$$

where Re refers to the real part of the complex number[4] and the dots above the symbols refer to the time derivative and indicate the harmonically time varying nature of these vector quantities. These are also referred to as phase vectors[5] with

---

[3] Refer to definition of the *Poynting Vector* later onwards in the text.

[4] The imaginary part represents the reactive power density in the near field and the real part the one in the far field which is of more interest here.

[5] In detail: $\dot{\vec{E}} = \hat{x}\, E_{x0}e^{j\omega t}$ with E_{x0} being the complex amplitude component in direction $\hat{x}$, j the imaginary unit, $\omega = 2\pi f$ the angular frequency, and t the time.

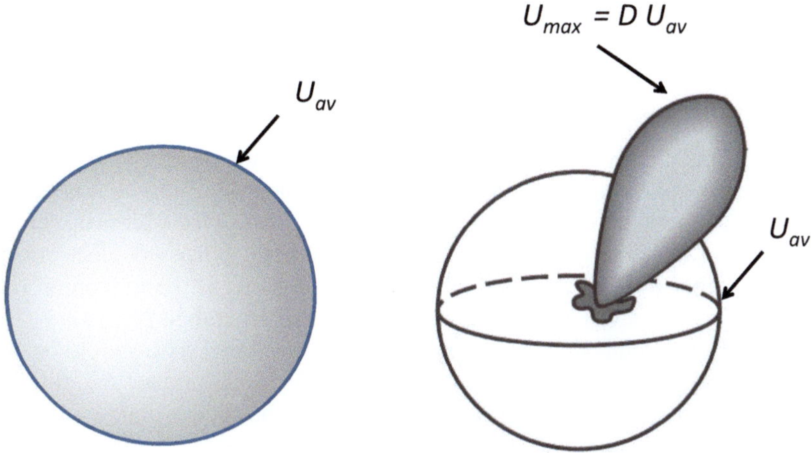

Fig. 5.6 Illustration of antenna directivity D, refer to text for more details

the asterisk designating to the complex conjugate quantity. The average *radiation power* P_{rad} through a closed surface around the radiation source can be obtained via

$$P_{rad} = \oint \int_A \vec{S}_{av} \cdot d\vec{A} = \frac{1}{2} \oint \int_A Re\left(\vec{E} \times \vec{H}^*\right) \cdot d\vec{A} \quad [\text{W}] \tag{5.5}$$

The radiation intensity provides the power per unit solid angle[6] and is given by $U(\Theta, \Phi) = \vec{S}_{av}\, r^2$ (W/unit solid angle). Based on the radiation intensity the antenna *directivity* D is defined as the ratio between maximum and average radiation intensity $D = U_{max}/U_{av}$ (see Fig. 5.6). Usually the quantity U_{av} is replaced by the radiation intensity of an ideal[7] isotropic source U_i having the same total radiation intensity given by $U_{av} = U_i = P_{rad}/4\pi$. In reality antennas will always generate radiation losses[8] which can be quantified by the dimensionless *antenna efficiency* k and the simple relation

$$G = \frac{k\,U_{max}(\Theta, \Phi)}{U_i} = k\,D \tag{5.6}$$

In case of a lossless antenna the gain equals its directivity ($k = 1$) otherwise it will be smaller ($k < 1$). For a parabolic dish antennas deployed in satellite TT&C

---

[6] The measure of solid angle is the *steradian* described as the solid angle with its vertex at the center of a sphere that has radius r.

[7] Ideal in this context means that no internal radiation losses exist.

[8] The two main sources of loss are heat losses due to the finite conductivity of the used antenna material and losses due to the dielectric antenna structure.

stations the antenna gain at the centre of the main beam can be derived from

$$G = \eta \, \frac{4 \, \pi \, A}{\lambda^2} \tag{5.7}$$

where A is the physical antenna area, λ the transmission or receiving wavelength, and η the so called aperture efficiency factor which considers a potential non-uniform illumination being typically in the range of 0.5–0.7 for microwave antennas.

As it is a characteristic of any antenna to radiate polarised electromagnetic waves,[9] another relevant antenna performance parameter is the maximum ratio of transmit power achieved in two orthogonal planes referred to as *axial ratio*. For the transmission of circularly polarised waves the desired axial ratio should be one (or zero, if expressed in deciBel[10]). A small axial ratio would reduce the polarisation losses if the transmitting and receiving waves do not have their planes of maximum gain aligned. If only a single plane of polarisation is needed, then high axial ratios are desired in order to discriminate against signals on the other polarisation.

5.3.1   The Receiving Path

The incoming signal path of a typical TT&C station is schematically depicted in Fig. 5.7. It starts from the incoming signal transmitted by the satellite RF transponder at the very left and ends with the demodulated TM frames that can be forwarded to the Satellite Control Facility (SCF) at the right side (refer to Chap. 7). For the sake of simplicity only the main building blocks of the RF chain are discussed here and components like switches, couplers, and filters which are also present are not further described here.

The incoming electromagnetic waves are collected by the beam-shaping antenna surface area and directed into the entrance window of the *feed system*. Depending on the antenna design this can either be located directly in the focal point of the parabolic dish or at the base of the reflector dish in the case of a Cassegrain design as shown in the upper panel of Fig. 5.4. The feed system itself consists of the *feed horn*, the *tracking coupler*, and the *diplexer* as shown in the sub-panel of Fig. 5.7. The *diplexer* allows to use the same feed system for both reception and transmission by separating the incoming and outgoing wave flows. After the Diplexer the signal is led via wave guides to the *low noise amplifier* (LNA) where its signal strength is amplified before being passed on to the *down-converter* (D/C). The D/C converts the frequency from its high frequency level (GHz) to a much lower level (typically MHz) which is referred to as the *intermediate frequency* (IF)[11]. The IF processor

---

[9] The tip of the electromagnetic wave vector remains in the plane of propagation for linearily polarised waves, whereas it rotates around the direction of propagation for circular or elliptically polarised waves.

[10] Ratio [dB] = $10 \log_{10}$ Ratio [physical unit].

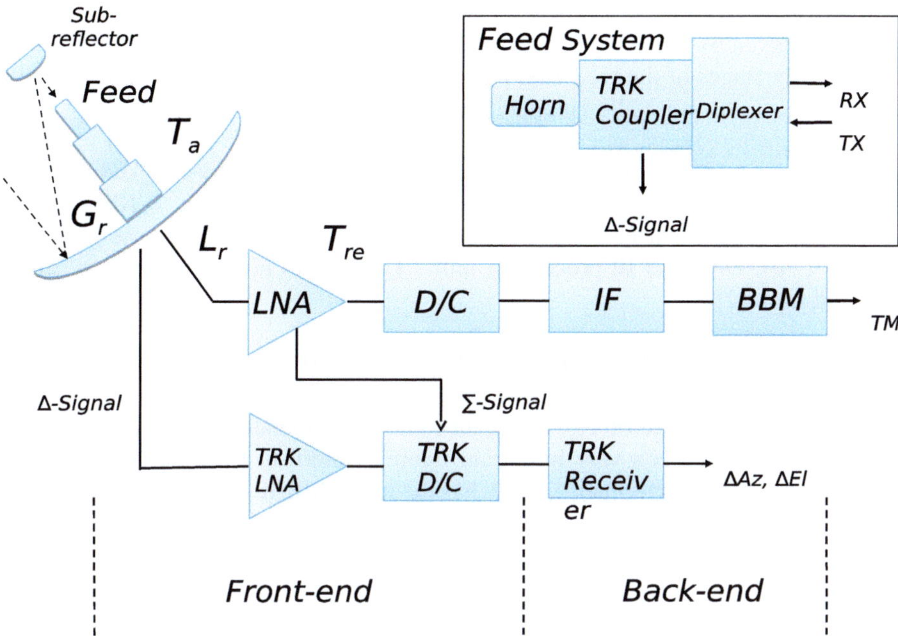

Fig. 5.7 Simplified incoming (receiving) signal path, starting from the left to the right; LNA = Low Noise Amplifier, D/C = Down-Converter, IF = Intermediate Frequency Processor, BBM = Baseband Modem, T_a = antenna noise temperature, G_r receiving antenna gain, L_r antenna feed to LNA waveguide loss

usually contains a *demultiplexer* to divide the output from the D/C into separate (frequency) channels. The last stage in this reception chain is the *baseband modem* (BBM) which demodulates the actual signal content (satellite telemetry or payload data) from the incoming signal.

The receiving performance of the TT&C antenna will determine the signal quality which can be defined as the received signal strength relative to the electrical noise of the signal background and is referred to as the *signal-to-noise-ratio* (SNR). The electrical noise stems from the random thermal motion of atoms and their electrons which can be considered as small currents that itself emit electromagnetic radiation. It can be shown that the noise N can be expressed as

$$N = k\,T\,B \tag{5.8}$$

where k is the Boltzmann constant ($k = 1.38 \times 10^{-23}$ J/K), T is the so called noise temperature in Kelvin, and B is the frequency bandwidth in Hz. The *power spectral*

---

[11] The down conversion from HF to IF is a very common process in electronic devices as signal processing is much easier done in a lower frequency range.

noise density N_0 is the noise power per unit of bandwidth expressed as W/Hz and therefore given by $N_0 = N/B$. This explains the reason to include band-pass filters at the early stages of the receiving path in order to reduce the bandwidth and with this the incoming noise allowing an immediate improvement of the SNR.

An characteristic parameter of every TT&C station is the combined receiving system noise temperature T_{sys} which takes into account the noise picked up by its antenna feed and reflector area T_a, the loss from the wave guide $l_r = 10^{L_r}/10$, and the receiver equivalent noise temperature T_{re} which considers the noise generated by the LNA, D/C, and IF processing units. The system noise temperature can be derived from [2]

$$T_{sys} = T_a/l_r + (1 - 1/l_r) \cdot 290 + T_{re} \tag{5.9}$$

With the knowledge of T_{sys}, the gain-to-noise-temperature ratio G/T can now be evaluated which is a very important figure of merit to characterise the TT&C station's receiving performance. G/T has the unit of dB/K and is defined as

$$\boxed{G/T = G_r - L_r - 10\log(T_{sys})} \tag{5.10}$$

5.3.2   *The Transmission Path*

The outgoing signal path is schematically shown in Fig. 5.8 which begins at the right side where the telecommand (TC) generated by the SCF is modulated onto the baseband signal by the BBM. After passing the IF processor, the conversion to high-frequency is done by the *up-converter* unit (U/C). Further filtering after this stage might be needed in order to reduce unwanted outputs. The next stage is the

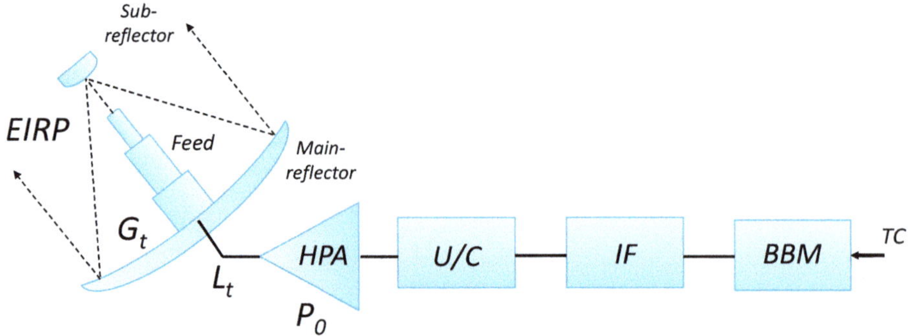

Fig. 5.8 Simplified outgoing (transmitting) signal path, starting from the right to the left. HPA = High Power Amplifier, U/C = Up-Converter, IF = Intermediate Frequency Processor, BBM = Baseband Modem

high power amplifier (HPA) whose main task is the signal power amplification to the target value P_0. As the HPA is a key component in every TT&C station, its design is explained in more detailed in the next section. After amplification the signal is directed through a wave guide or coaxial cable connection to the antenna feed. The loss that occurs on this path is referred to as the *transmission line loss L_t* and is an important parameter to be considered in the evaluation of the overall *link budget.* From the feed the signal is radiated into space using the antenna reflecting area and its geometry that determines the beam shaping (and focusing) ability expressed as the antenna gain value G_t. The three parameters P_0, L_t, and G_t determine the transmit RF performance referred to as *effective isotropic radiated power* or $EIRP$ and is expressed in dBW.[12] The EIRP can be derived from the following power balance equation

$$\boxed{EIRP = P_0 - L_t + G_t}$$

(5.11)

5.3.3   High Power Amplifier Design

The design of a high power amplifier (HPA) needs to consider the frequency and bandwidth range it has to operate with, the required gain of power of its output (including the linearity over the entire frequency range), the efficiency (power consumption), and finally its reliability. There are three generic types of HPAs in use today, the *Klystron power amplifier* (KPA), the *travelling wave tube amplifier* (TWTA), and the *solid state power amplifier* (SSPA).

Both KPA and TWTA implement an electron gun which generates an electron beam inside a vacuum tube by heating up a metal cathode and emitting electrons via the thermoionic emission effect.[13] As the cathode and anode are put at different potentials, the electron beam is accelerated through the vacuum tube into the direction of the collector (see upper panel of Fig. 5.9). Further beam shaping can be achieved by focussing magnets. Inside the same tube a helical coil is placed to which the RF signal is coupled either via coaxial or waveguide couplers. The helix is able to reduce the phase velocity of the RF signal and generate a so called slow wave. If correctly dimensioned, the electron velocity (determined by the length of the tube and difference of potential) and the wave phase velocity (determined by the diameter of the helix coil and its length) can be synchronised. In this case the electrons form clusters that can be understood as areas of higher electron population density which replicate the incoming RF waveform. These clusters then induce a new waveform into the helix having the same time dependency (frequency) as the input RF signal but a much higher energy. This amplified signal is then lead to the

---

[12] This can be understood as the power gain expressed relative to one Watt.

[13] Heat is transferred into kinetic energy of electrons which allows them to leave their hosting atoms and produce an electron cloud that surrounding the heated metal.

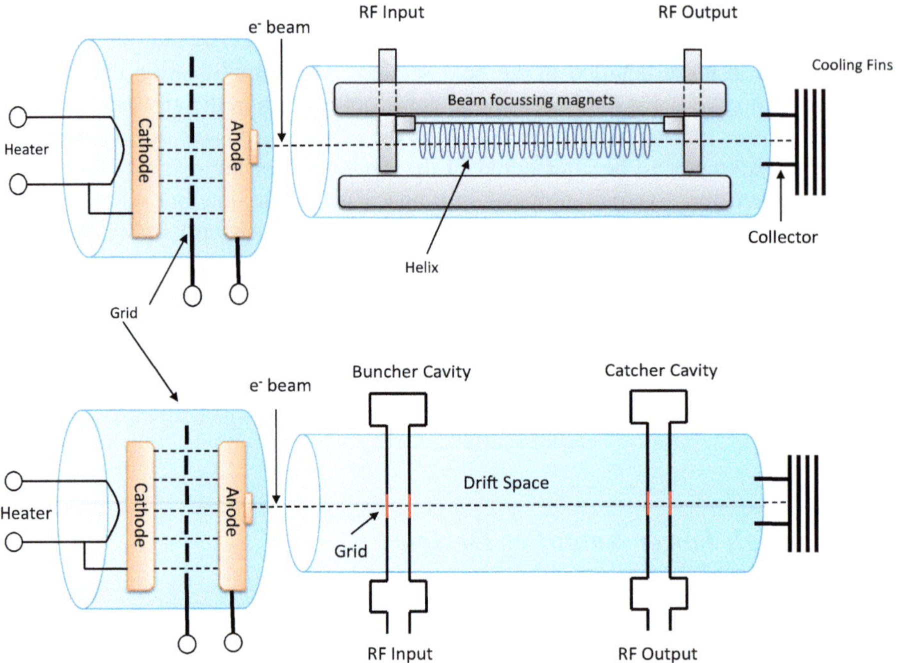

Fig. 5.9 Principal design of a travelling wave tube (upper panel) and a Klystron power amplifier (lower panel)

RF output. The achieved amplification can be quite significant and reach an order of magnitude in the range of 40 to 60 dB, which refers to a factor of 10^3 to 10^6 times the input power.

The Klystron power amplifier (KPA) design is schematically shown in the lower panel of Fig. 5.9 and implements a set of resonant cavities (typically five, but only two are shown for simplicity here) into which the RF wave is coupled to generate a standing wave. These cavities have a grid attached to each side through which the electron beam passes. Depending on the oscillating electrical field, the electrons get either accelerated or decelerated and are therefore velocity modulated. This results in a density modulation further down the tube which is also referred to as electron bunching with the optimum bunching achieved in the output (or catcher) cavity. The density modulated electron beam itself generates an output signal with the required amplification. The gain can be up to 15 dB per cavity therefore achieving in sum up to 75 dB amplification with 5 cavities.

The SSPA does not involve a vacuum tube but uses semiconductor based field-effect transistors (FET) for the power amplification. The commonly used semiconductor material is Gallium Arsenide and the devices are therefore referred to as GaAsFET. As the maximum output of a singe transistor is limited, transistors are combined to form modules. Even higher powers are achieved by combining several modules.

From an operational point of view it can be said that KPAs are narrow band devices with relative high efficiency and stability whereas TWTAs and SSPAs can be used for wide band applications. An important HPA characteristic is linearity which refers to the transfer characteristic between input and output power following a straight line over its operating frequency range.[14] Non-linearity can lead to the well known intermodulation interference which can cause disturbance among the various channels if the FDMA access protocol[15] is used. In this respect the SSPA usually reaches better linearity performances compared to TWTA and KPA.

5.3.4   Monopulse Tracking

Most TT&C antenna systems today support the autotrack mode that allows them to autonomously follow a satellite once a stable RF link has been established. This functionality can be realised via different methods and a widely used one is the *monopulse* technique which will be described in more detail here.[16] The starting point of monopulse is the *tracking coupler* which is deployed in the Feed System and is able to build the difference of two incoming beams and extract this as the so called Δ-signal (see Fig. 5.7). The two beams usually are the two polarisations of an incoming RF signal. This Δ-signal is guided to a tracking low noise amplifier (TRK/LNA) for amplification, followed by its down-conversion to IF by a tracking down-converter (TRK-D/C). Using the Δ- and the Σ-signal (i.e., the sum of the two beams available in the receiving path after the LNA), the *tracking receiver* derives the required corrections in azimuth and elevation in order to re-point the antenna to the source.

5.3.5   The Link Budget

The design of a reliable communications link between a sender and a receiver requires the evaluation of a representative link budget that properly considers the signal source, its amplification, and all the signal losses that occur on the transmission or reception path. Based on the definition of the radiated power or

---

[14] This can be expressed as maximum allowed gain slope requirement.

[15] Frequency Division Multiple Access means that several users can access the same transponder (or carrier) as they are separated in their sub-carrier frequency (referred to as channel).

[16] An alternative method used is the *conical scan method* in which the antenna feed performs a slight rotation around the signal emitting source. This leads to an amplitude oscillation of the received signal which is analysed and used to steer the antenna and centre it onto the source.

EIRP at the sending end and the gain-to-noise temperature G/T at the receiving end, the link budget can be computed as the ratio of (received) carrier power C to spectral noise level N_0

$$\frac{C}{N} = \frac{C}{N_0 B} = EIRP - L_{space} - L_{atm} + G/T - 10\log k \qquad (5.12)$$

where $L_{space} = (4\pi d f/c)^2$ is the *free space loss* that considers the signal attenuation due to the propagation loss over the travelled distance d. The frequency dependency of L_{space} is due to the fact that it is defined with respect to an isotropic antenna gain at the receiving end. L_{atm} is the signal attenuation caused by the atmosphere and k is Boltzmann's constant ($10\log k = -228.6$ dBW/K/Hz). For the Earth the main constituents causing atmospheric loss are oxygen, nitrogen, water vapour, and rain precipitation. It is important to note that signal loss due to rain can be significant for frequencies above 12 GHz as the corresponding wavelengths have similar dimensions to the rain drops which makes radiation scattering the dominant effect.

The final quality of a received signal depends on both, the carrier-to-noise ratio at RF level and the signal-to-noise ratio at baseband level where the actual demodulation occurs. At baseband level the type of signal modulation or *channel coding* applied to the data prior to transmission plays a fundamental role. Coding schemes are able to improve the transmission capability of a link allowing to achieve a higher data throughput with lower transmission errors for a given signal strength. An important performance measure of any transmission channel therefore is the received signal energy per bit E_b which has to exceed a minimum threshold in order to guarantee an error free signal decoding at baseband level. In other words, the probability P_e of an error in decoding any bit (referred to as the *bit error rate* or BER) decreases with an increase of the received *bit* signal-to-noise ratio E_b/N_0 (bit-SNR). This is demonstrated in Fig. 5.10 which compares the performance of a set of channel coding schemes as defined by the applicable standard CCSDS-130.1-G-3 [4] by showing their (simulated) BER as a function of received bit-SNR. The same relationship is also shown for an uncoded channel for comparison and clearly proofs the improvement in BER achieved by all coding schemes. Figure 5.10 can be useful to dimension a transmission channel as it allows to find for a given channel coding scheme the necessary minimum E_b/N_0 at signal reception in order to not exceed the maximum allowed BER.[17] If the coding scheme and BER have been fixed, the required minimum E_b/N_0 value determines the corresponding C/N_0 which is then an input to the link budget equation in order to solve the remaining system design parameters.

---

[17] Channel coding and decoding algorithms are usually part of the BBM functionality which also involves the application of various error correction techniques able to correct or compensate lost information due to low signal quality.

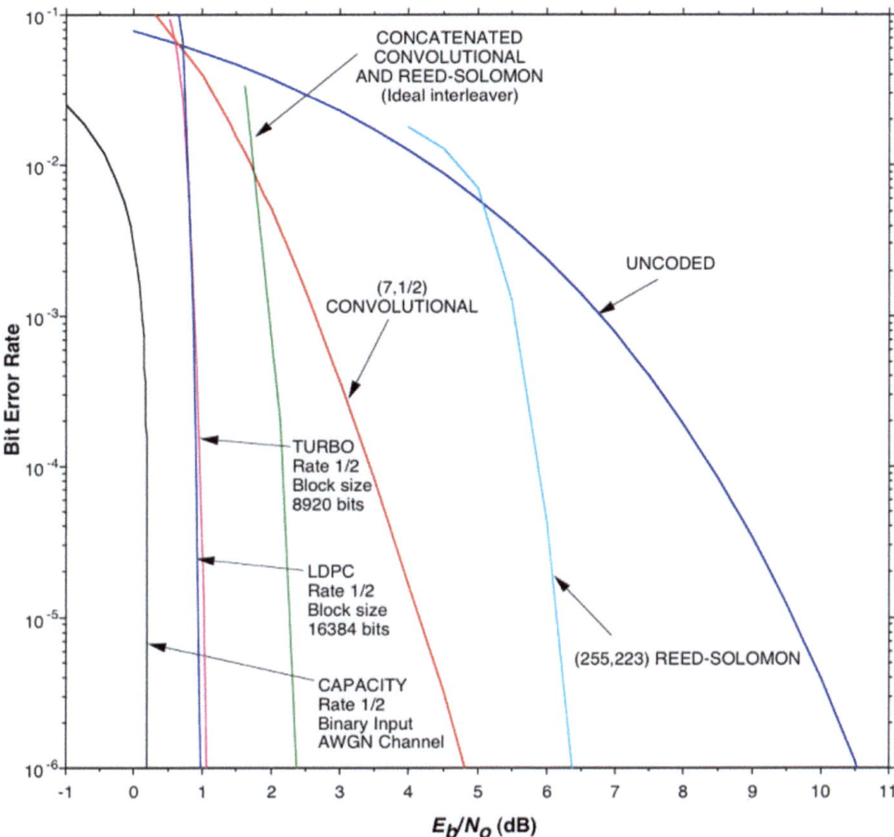

Fig. 5.10 Performance comparison of CCSDS recommended channel codes with an uncoded channel (blue line) and the lowest possible performance of a 1/2 rate code as given by the code-rate-dependent Shannon-limit [3] (line labelled CAPACITY Rate 1/2). LDPC = Low Density Parity Check; Reprinted from Figs. 3–5 of CCSDS-130.1-G-3 [4] with permission of the Consultative Committee for Space Data Systems

It is important to note that the link budget equation needs to be solved in two directions, once for the transmission from ground to space for which the required EIRP needs to be provided by the TT&C station, taking into account the G/T of the satellite transponder, and second for the transmissions from space to ground where the required EIRP needs to be achieved by the satellite transponder (and the satellite HPA) taking into account the ground station G/T. In case of a shortcoming of EIRP at satellite level, some compensations on ground can be achieved by increasing the receiver gain through a larger antenna size or shortening the wave guide distance between antenna and LNA which would both improve the antenna G/T.

5.3.6   *Frequency Compensation*

To avoid interferences between the signal transmitted by the TT&C ground station (uplink) and the one received from the satellite (downlink), different frequencies are usually used for each direction. The uplink frequency as transmitted by the ground station (shown as $f_{up,tr}$ in Fig. 5.11) will be Doppler shifted due to the relative radial velocity $\dot{\rho}$ (range rate) between the satellite and the TT&C station which is usually defined as positive in case the slant range ρ increases and negative otherwise. The uplink frequency as seen by the satellite transponder is therefore Doppler shifted and can be derived from

$$f_{up,rc} = f_{up,tr}\sqrt{\frac{1 - \dot{\rho}/c}{1 + \dot{\rho}/c}} \tag{5.13}$$

As the the satellite transponder is designed (and optimised) to lock at the nominal uplink centre frequency f_0, the station uplink frequency needs to be corrected by the expected Doppler shift $\Delta f_{Doppler} = f_{up,rc} - f_{up,tr}$ which can be computed on ground based on the predicted satellite orbit. Due to the satellite orbital motion v_{orb} and the Earth's movement v_{Earth} as indicated in Fig. 5.11, the Doppler shift is a

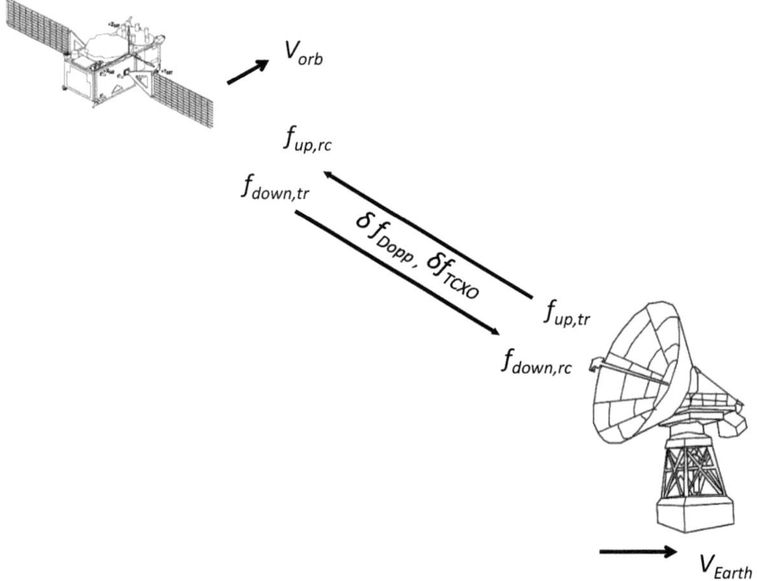

Fig. 5.11 Uplink and downlink frequency adjustments. v_{orb} = satellite orbit velocity, $f_{up,tr}$ = transmitted uplink frequency, $f_{up,rc}$ = received uplink frequency, $f_{down,tr}$ = transmitted downlink frequency, $f_{down,rc}$ = received downlink frequency, δf_{Dopp} = frequency shift introduced by Doppler, δf_{TCXO} = frequency shift due to satellite oscillator drift

dynamic value that continuously changes throughout a satellite pass. A prediction of the expected Doppler shift profile $\Delta f_{Doppler}(t)$ specific for each contact geometry is therefore required and is referred to as a Doppler pre-compensation file.

On downlink the satellite transponder transmits at frequency $f_{down,tr}$ which again is Doppler shifted by $\Delta f_{Doppler}(t)$. In addition the satellite transponder's internal oscillator (usually realised as a temperature controlled crystal oscillators or TCXO) can change its reference frequency due to ageing or uncompensated temperature drift which causes an additional frequency shift Δf_{TCXO} of the downlink frequency $f_{down,rc}$. In contrast to the Doppler, Δf_{TCXO} can however not be predicted but only measured on ground. When the BBM demodulator achieves lock, $f_{down,rc}$ can be measured and Δf_{TCXO} derived from

$$\Delta f_{TCXO} = f_{down,rc} - f_{down,tr} + \Delta f_{Doppler}(t) \tag{5.14}$$

The TCXO drift should be estimated at every contact and stored in a specific ground station database so it can be used for subsequent contacts.

There are usually two different modes of satellite transponder operation. The first one is referred to as *non-coherent mode* in which $f_{down,tr}$ is a predetermined frequency. As this frequency is however affected by internal oscillator drifts, the ground station demodulator needs to perform a search before being able to lock on the signal. In *coherent* mode the satellite transponder sets its downlink frequency to a value derived from the uplink frequency and multiplied by a known constant factor (typically expressed as a ratio of two numbers) which allows to ignore Δf_{TCXO}.

5.4   Remote Sites

5.4.1   Selection and Preparation

The choice of an adequate remote site for the installation of a new TT&C ground station is a complex process and needs to consider a number of criteria (or even requirements) that need to be fulfilled in order to guarantee its reliable operation and maintenance. The list below provide some important points that need to be considered as part of the site selection but should by no means seen as an exhaustive list as there are usually also non-technical criteria (e.g., geopolitical relevance of a territory, contractual aspects, security, etc.) that are not addressed here.

- The geographical location of TT&C station can have a major impact on the possible visibilities and derived contact duration between a satellite and the station. Especially the location latitude (e.g., polar versus equatorial) should be a relevant selection criteria as it can significantly influence the station's usability for a satellite project.
- Site accessibility (e.g., via airports, railways, roads etc.) plays an important role during both the construction phase and the maintenance phase of a ground

station. Especially during the construction phase, large items (potentially packed in overseas containers) will have to be transported to site and depend on the availability of proper roads in terms of size and load capacity in order to reach the construction site. People will have to travel to site during both the construction and the maintenance phases in order to perform installation or regular repair activities.

- In order to guarantee a low RF noise environment, TT&C antennas are often built at locations that are either far away from noise sources (e.g., mobile telephone phone mast, radio or television transmitter masts, or simply densely populated areas) or they are built on locations where the natural terrain can serve as an RF shielding like distant mountains. This however should not be in conflict with requirements of a free horizon profile which demands to avoid high elevation obstacles or the need for a good site accessibility as described in the bullet above. An appropriate balance needs to be found here.
- In order to host a TT&C antenna, a remote site needs to provide the necessary infrastructure for power, telephone, and a Wide Area Network (WAN) that allows a stable connection to the remaining ground segment potentially thousands of kilometres away. For a completely new site with no such infrastructure in place yet, the deployment time and cost needs to be properly estimated and considered in the selection process. For the power provision there are typically specific requirements related to stability which are described in more detail below.
- The antenna foundations require a sufficient soil bearing capacity in order to avoid any unwanted soil movement during the deployment and operational phase. Soil stability is not only relevant for the antenna foundation but also for the surrounding ground which needs to support heavy crane lifting during the construction work. To guarantee this, dedicated soil tests should be performed as part of the site inspection and selection phases and if needed appropriate measure need to be taken to reinforce the soil (e.g., the deployment of pillars deep into the ground that are able to reach a region of more stable soil properties (e.g., rock).
- Severe weather conditions like snow, heavy rain with floods, strong winds, or even earthquakes need to be identified from historic weather and climate statistics and, where applicable, considered with appropriate margins to account for future climate warming effects. The antenna design needs to consider appropriate sensors and warning techniques and deploy mitigation measures (e.g., drainage channels) to counteract such extreme conditions.

The site preparation phase needs to ensure the provision of all relevant site infrastructure required for the installation and proper operation of the TT&C ground station. The majority of the related activities will typically have to precede the actual station deployment but some activities can potentially be performed in parallel. Depending on the industrial set up of a project, the site preparation and actual station build activities could very well be covered by separate contractual agreements and different subcontractors. In this case a careful coordination of the activities need to be ensured both from a timing and a technical interface perspective point of view.

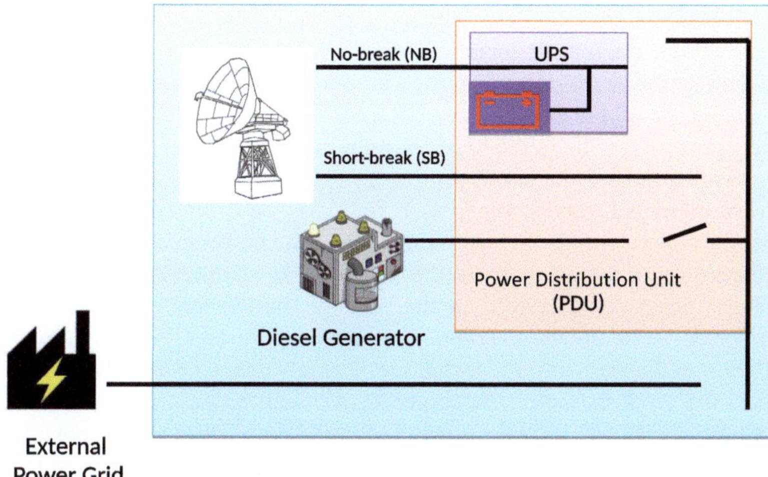

Fig. 5.12 Schematic view of power sources and distribution. UPS = Uninterruptible Power Supply

5.4.2 *Power: No-break and Short-break*

A very important site preparation activity is the deployment of all relevant equipment to supply and distribute the required power to operate the station. The *power distribution unit* (see Fig. 5.12) is the central power management unit in the station whos main task is to distribute the power via the *short-break* and the *no-break* connections to the various subsystems. As implied by the naming convention, the no-break supplies are designed to provide uninterrupted power supply even if in case of a service interruption of the external power grid. In order to implement such a service the PCDU needs to manage the following types of power sources and be able to switch between them:

- A long-term power source should be available for the day-to-day station operation and is usually provided via an external public power grid. As with any public power supply there is no guarantee for uninterrupted service provision. Even if used as the principal power source backup sources need to be put in place that can bridge a power outage for limited amount of time.
- A medium-term power source that can be realised via a small and local power plant and is directly deployed on-site. Its deployment, operation, and regular maintenance falls under the control and responsibility of the site hosting entity, which makes it an independent power source from external factors. The most common realisation of such a local plant is a Diesel generator (DG) that has low complexity and is therefore rather easy to maintain and has high reliability. The deployment of a DG needs to also consider all the infrastructure required to store and refill fuel and at the same time respect all the applicable safety regulations.

The operational cost of a DG is mainly driven by the cost of its fuel which can be quite high for remote locations which makes this power source only viable for short term use in order to bridge an outage of the external power grid. The dimensioning of the fuel storage determines the maximum possible length of power provision without the need for refuelling and should cover for a minimum of several weeks of operations.

- A short-term but very stable power source is an essential equipment for any critical infrastructure as it can provide immediate power in case of an interruption of the principal source without the need of any start-up or warm-up time like a Diesel generator described in the previous bullet. Such a short-term source is referred to as an *uninterruptible power supply* or UPS and implements a battery based energy storage technique. The advantage of immediate availability makes the UPS the preferred choice to provide the so called *no-break* power supplies in a TT&C station to which all critical components are connected that cannot afford any power interruption. The main disadvantage of a UPS is its very limited operational time span due to the limited power storage capacity of today's batteries[18] and the quite demanding power needs of the various TT&C subsystems which is also the reason why not all components will usually be connected to the no-break supplies.

5.4.3 Lightning Protection

TT&C ground stations are composed of large metallic structures with the most dominant ones being the large parabolic antenna mounted on top of the antenna building. As such dishes can have diameters up to 30 meters if dimensioned for deep space communication. During operations they will move several meters above the ground and can therefore attract lightning. The deployment of adequate lightning protection is therefore of paramount importance to protect people and equipment and needs to be considered already during the site preparation phase. The sizing of the lightning system comprises the evaluation (and subsequent deployment) of the required number of lightning masts, their required (minimum) heights, as well as their optimum position. In addition the proper implementation of the grounding system needs to be considered early onwards in the process as this will usually require soil movement.

5.4.4 Site Security

As the TT&C station is a crucial asset for the commanding and control of the space segment, its protection against unauthorised access is an important aspect that needs

---

[18] A typical UPS deployed in a TT&C station can usually bridge only up to a few hours of operations.

to be considered during the site preparation phase. The necessary equipment is determined by the applicable security requirements of the project. Typical assets are fences with additional optical barriers (e.g., infrared sensors), guard houses, intrusion detectors and their connection to an alarm monitoring system. Also firewalls to protect against cyber attacks need to be considered for relevant network connections. Finally physical fire detection and protection systems need to be put in place which should conform with the local safety regulations.

5.5 Interfaces

The most important interfaces between the TT&C and the other GCS elements are shown in Fig. 5.13 and described in the bullets below:

- **FDF** (see Chap. 6): in order for a TT&C station to find a satellite and establish a contact, it requires the up-to-date orbit predictions from the Flight Dynamics Facility. These can be either provided as an orbit prediction file (e.g., a set of position vectors as function of time), a Two-Line-Element (TLE), or a set of spherical coordinates (e.g. azimuth and elevation as function of time). These files are also referred to as *pointing files* as they serve the ground station to point to the correct direction in the sky, either as an initial aid before the autotrack function of the antenna can take over, or as a continuous profile for the antenna control unit to follow during the execution of a pass (program track). An important aspect

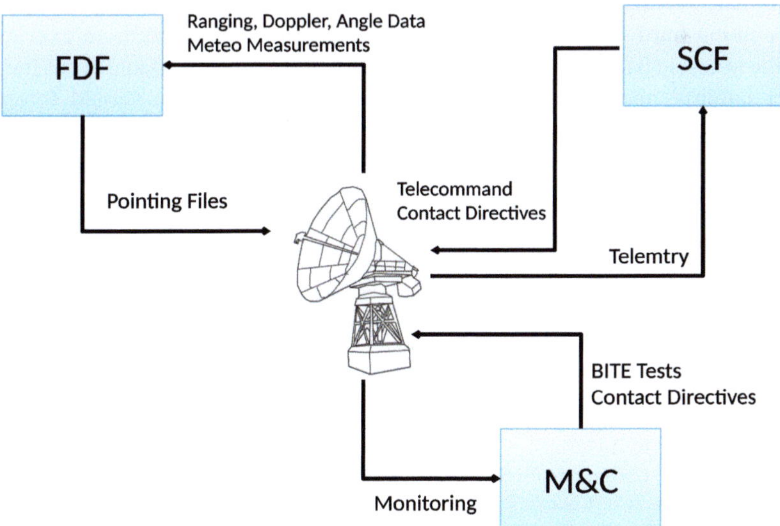

Fig. 5.13 TT&C interfaces to other GCS elements: SCF = Satellite Control Facility, M&C = Monitoring and Control Facility, FDF = Flight Dynamics Facility

here is the time span of validity of the pointing files which needs to be checked prior to each contact in order to avoid the loss of a contact due to an inaccurate input.

As part of nominal contact operations the TT&C has to acquire radiometric data (i.e., range, Doppler, and angular measurements) from the satellite which need to be sent to FDF as input for orbit determination. In addition, meteorological measurements should be taken prior or during each contact/pass and provided to FDF as an input for the atmospheric signal delay modelling in the orbit determination process.

- **SCF** (see Chap. 7): to be able to perform automated satellite operations, the Satellite Control Facility should be able to initiate a satellite contact via a set of commands referred to as *contact directives*. These commands should establish the applicable data transfer protocol (e.g., CCSDS SLE), perform the correct configuration of the RF transmit path, and finally raise the carrier to start the actual transmission of TCs to the satellite. Once the satellite transmits its telemtry (TM) to the TT&C station and the BBMs have demodulated the data from the RF signal, the TM needs to be forwarded to the SCF for further analysis and archiving.

- **MCF** (see Chap. 10): the interface to the Monitoring and Control Facility enables the ground segment operator in the control centre to maintain an up-to-date knowledge of the operational and health status of the entire TT&C station. It is also important to access to the most relevant subsystem parameters in order to gain a better understanding of potential anomalies. For the most critical components of the TT&C station (e.g., BBMs, LNA, U/C, D/C, HPAs), the station design should foresee an adequate level of redundancy and implement automated failure detection and fail-over mechanisms. In order to find potential upcoming hardware failures at an early stage, the regular remote execution of built-in tests (BITE) should be performed which can be initiated from the MCF. For failures, anomalies, and warnings the interface design should foresee the transmission of corresponding messages to the M&C facility.

References

1. Vertex GmbH. (2013). S-Band 13m TT&C Galileo. In *Installation manual mechanical subsystem*. GALF-MAN-VA-GCBA/00002, Release 1.0.
2. Elbert, B. (2014). *The satellite communication ground segment and earth station handbook* (2nd ed.). Norwood: Artech House.
3. Shannon, C. E. (1948). A mathematical theory of communication. *Bell System Technical Journal, 27*(3), 379–423.
4. The Consultative Committee for Space Data Systems. (2020). *TM Synchronization and Channel Coding — Summary of Concept and Rationale*. CCSDS 130.1-G-3. Cambridge: Green Book.

Chapter 6
The Flight Dynamics Facility

The Flight Dynamics Facility (FDF) can be considered as the element with the highest incorporation of complex mathematical algorithms and data analysis techniques. It therefore requires a lot of effort to ensure a thorough validation of its computational output, especially for parameters that are commanded to the satellite. Flaws in complex algorithms might not always be immediately evident to the user but can still lead to serious problems during operations. Especially for computational software that is developed from scratch and lacks any previous flight experience, the cross-validation with external software of similar (trusted) facilities is highly recommended. The FDF has a very large range of important computational tasks of which the most relevant ones are briefly summarised in the following bullets:

- The regular orbit determination (OD) is one of the central responsibilities of the FDF and requires as input radiometric tracking data measured by the TT&C stations. In other words, the FDF can be considered as the element that is responsible to maintain the up-to-date operational orbit knowledge and to derive all relevant products from it.
- One such orbit product is the pointing file required by the antenna control unit of the TT&C station (see Chap. 5), allowing it to direct the antenna dish and its feed to the correct location on the sky where the satellite is expected at the start of an upcoming pass. An inaccurate or incorrect orbit determination would also imply inaccurate pointing files that could imply the loss of the pass. The lack of newly gained radiometric measurements would then further deteriorate the satellite orbit knowledge.
- Not only the TT&C station requires up-to-date orbit knowledge but also the satellite itself in order to perform future onboard orbit predictions and attitude control. This is done through the regular transmission of up-to-date orbit elements in the form of the onboard orbit propagator (OOP) TC parameters which are prepared by the FDF (see also Sect. 6.6).

© The Author(s), under exclusive license to Springer Nature Switzerland AG 2023
B. Nejad, *Introduction to Satellite Ground Segment Systems Engineering*, Space
Technology Library 41, https://doi.org/10.1007/978-3-031-15900-8_6

- Using the most recent orbit knowledge as initial conditions, the FDF performs orbit predictions from a given start epoch into the future using a realistic force model that is adequate for the specific orbit region. Whereas satellites orbiting at altitudes lower than about 1000 km require the proper modelling of atmospheric drag as one of the dominant perturbing forces, orbits at higher altitudes (MEO and GEO) will be mainly influenced by luni-solar forces and solar radiation pressure and atmospheric drag becomes negligible. Orbit predictions are an important means to assess the future development of the orbital elements and detect potential violations of station keeping requirements. Orbit predictions are also required to compute orbital events like eclipses, node crossings, TT&C station visibilities, or sensor field-of-view crossings to name only the most typical examples.
- The determination of the satellite's attitude profile is a similar task to orbit determination but based on the inputs from the satellite's attitude sensors (e.g., star trackers unit, gyros, Sun sensors, Earth IR sensor, etc.). Attitude *prediction* in contrast does not require any external sensor inputs but uses a predefined attitude model to compute a profile that can either be as expressed as a set of three angular values (e.g., raw, pitch, and yaw) or a 4-dimensional quaternion.
- Most satellites will be subject to station keeping requirements which mandate them to remain within a given box around a reference location on their orbit. This requires the FDF to perform long-term orbit predictions in order to detect a potential boundary violation. The time of such a violation is an important input to the manoeuvre planning module that is supposed to compute an orbit correction (station keeping) manoeuvre to take the satellite back inside its control box.
- The space debris population has unfortunately increased significantly during the past decade and poses today a significant risk to operational satellites deployed in almost all orbit regions. This makes a close and continuous monitoring of the space environment an essential task for every space project. Having the best orbit knowledge of the own satellite, the FDF can use an orbit catalogue of known space debris as an input to perform a collision risk analysis. In case a non negligible collision risk has been identified, adequate avoidance manoeuvres need to be computed and commanded to the satellite.

6.1 Architecture

The functional architecture of the FDF can be derived from the main tasks described in the preceding section and is schematically depicted in in Fig. 6.1.

Both the orbit and attitude determination components make use of similar core algorithm that implement some kind of least-squares or Kalman batch filter which is described in more detail in Sect. 6.3. The relevant input data for these modules are the radiometric tracking data from the TT&C station network and attitude sensor measurements obtained from satellite telemetry.

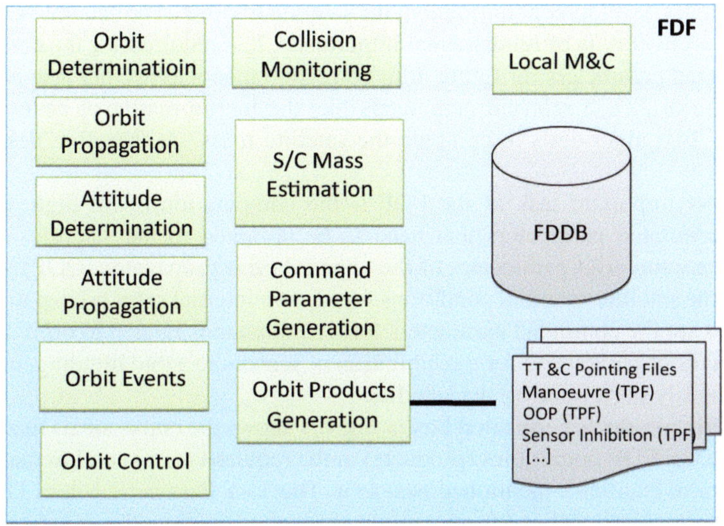

Fig. 6.1 High level architecture of the Flight Dynamics Facility (FDF)

The orbit propagation function implements mathematical routines for the numerical integration of the equations of motion and models the relevant external forces acting on the satellite. The most relevant forces are the Earth's gravity field and its perturbations due to the planet's oblateness, atmospheric drag, third body perturbations, and solar radiation pressure. Precise orbit determination needs to consider even more subtle forces that are caused by the Earth radiation pressure, tides, relativistic effects, and extremely weak forces that can stem from material properties and their imbalanced surface heating.[1]

The orbit events function is responsible for the computation of all events relevant for the operations of the satellites. Apart from generic events like node crossings or eclipses, there are also satellite platform specific ones, like sensor blindings or FoV crossings which are defined and described in the relevant satellite user manuals. The timing of such events is an important input for the operational planning of satellite activities like the change of attitude modes (e.g., from Sun to Earth nadir pointing) or the execution of orbit correction manoeuvres which could require the presence of the Sun in specific attitude sensor field-of-view.

The orbit control module implements algorithms for the computation of orbit corrections both in terms of Δv and the required thrust orientation. This module requires as input the up-to-date orbit position from the orbit determination module.

After the execution of every orbit correction manoeuvre, the satellite mass will decrease due to the fuel used by thruster activity. A dedicated module is therefore required to compute the consumed and remaining fuel and keep a record of the

---

[1] This type of forces are also referred to as *empirical forces*.

satellite mass development throughout the satellite lifetime. The accurate knowledge of the mass history is of fundamental importance. It is required for the computation of command parameters for future orbit correction manoeuvres, but also needed for the update of AOCS relevant parameters like the inertia matrix or centre of mass position. Two algorithms to compute the satellite mass are therefore described in Sect. 6.5.

Another important task of the FDF is the computation of all flight dynamics related command parameters that need to be uplinked to the satellite. A typical example are the orbit parameters of the onboard orbit propagator (OOP) which are used by the satellite as initial conditions for its own internal orbit propagator.[2] Other examples are the command parameters for telecommands related to orbit correction manoeuvres or parameters for the inhibition of sensors to avoid blinding, in case not autonomously performed by the satellite.

The orbit products generation box in Fig. 6.1 shows the components that generate the full set of FDF output files (products) in the required format and frequency with a few typical examples mentioned next to it. The Task Parameter Files (TPFs) is an example of a clearly defined format used to exchange command parameters with the Satellite Control Facility (SCF) (cf. [1])

The FDDB in Fig. 6.1 refers to the *flight dynamics database* which contains all the necessary parameters of the satellite and the orbit environment that are used by the various algorithms implemented in the FDF with a few examples listed in Table 6.1.

6.2   Orbit Propagation

Orbit propagation is performed in order to advance orbit information from one initial epoch t_0 to a target epoch t_1. The target epoch will usually lie somewhere in the future but could in theory also be in the past. The following fundamental properties are worth mentioning in relation to orbit propagation:

- The accuracy of the propagated orbit is defined by the quality of the initial orbital elements at t_0, the quality of the applied force model, and assumptions in the propagation algorithm (e.g., step size, applied interpolation techniques, etc.);
- The accuracy of the propagated orbit degrades with the length of the propagation interval. This is mainly caused by the build up of truncation and round-off errors due to a fixed computer word length.

Orbit propagation techniques need to solve the equations of motion of the flying satellite in orbit. From Newton's second law the equation of motion can be simply written as

---

[2] The OOP parameters computed on ground are usually uplinked via a dedicated TC around once per week. The exact frequency however depends on the complexity of the OOP algorithm implemented in the satellite onboard software.

Table 6.1 Summary of important parameters in the Flight Dynamics Database (FDDB)

FDDB parameter	Description
Mass properties	Satellite dry and wet mass (the wet mass value is usually only available after the tank filling has been completed at the launch site).
Inertia tensor	This includes relevant information for interpolation due to a change of mass properties.
Centre of mass (CoM)	Location (body fixed frame) at initial launch configuration and predicted evolution (e.g., interpolation curves as function of filling ratio).
Attitude	Position and orientation of all attitude sensors (e.g., body frame mounting matrix); Field-of-View (FoV) dimensions and orientation.
Thruster	Mounting position and orientation, thrust force and mass flow (function of tank pressure), efficiency factor, etc.
Propellant	Parameters required for fuel estimation (e.g., hydrazine density, tank volume at reference pressure, filling ratio, etc.)
Planetary orientation	For Earth Orientation Parameters(EOP) refer to relevant IERS publications [2].
Time system	Parameters (e.g., leap seconds) required for the transformation between time systems (UT1, UTC, GST).
Gravity field	Parameters for the definition of the central body gravitational field.
Physical constants	Relevant physical and planetary constants required by the various FD algorithms.

$$\sum \vec{F} = \frac{d(m\,\dot{\vec{r}})}{dt} = m\,\ddot{\vec{r}} \tag{6.1}$$

where $\vec{r}$ is the position vector of the satellite expressed in an Earth centred inertial coordinate system referred to as ECI (see Appendix A),[3] $\ddot{\vec{r}}$ is the corresponding double time derivative of the position vector (acceleration), m the satellite mass, and $\vec{F}$ is the force vector acting on the satellite. If one combines the satellite mass and the mass of the central body M into the term $\mu = G(m + M)$, the equation of (relative) motion can be expressed in the so called *Cowell's formulation*

$$\ddot{\vec{r}} = -\frac{\mu \vec{r}}{r^3} + \vec{f}_p \tag{6.2}$$

where the first term on the right hand side represents the central force and $\vec{f}_p$ the perturbing forces acting on the satellite. The latter one can be further decomposed into several components

$$\vec{f}_p = \vec{f}_{NS} + \vec{f}_{3B} + \vec{f}_g + \vec{f}_{Drag} + \vec{f}_{SRP} + \vec{f}_{ERP} + \vec{f}_{other} \tag{6.3}$$

---

[3] In a rotating coordinate system a more complex formulation is needed that also considers expressions for virtual force components like the Coriolis force.

Table 6.2 Summary of perturbing forces in Eq. (6.2). The mathematical expressions for the various for contributions were adapted from [3]

Force	Description	Mathematical expression
$\vec{f}_{NS}$	Non-spherical gravitational influence including tidal contributions	$\vec{f}_{NS} = T_{xyz}^{XYZ} T_{r\phi\lambda}^{xyz} \nabla U'$
$\vec{f}_{3B}$	Third-body effects	$\vec{f}_{3B} = \sum_{j=1}^{n_p} \mu_j \left(\frac{\vec{\Delta}}{\Delta_j^3} - \frac{\vec{r}_j}{r_j^3} \right)$
$\vec{f}_g$	General relativity contribution	$\vec{f}_g = \frac{\mu}{c^2 r^3} \left\{ \left[2(\beta + \gamma)\frac{\mu}{r} - \gamma(\vec{r}\cdot\vec{r}) \right] \vec{r} + 2(1+\gamma)(\vec{r}\cdot\dot{\vec{r}})\dot{\vec{r}} \right\}$
$\vec{f}_{Drag}$	Atmospheric drag	$\vec{f}_{Drag} = -\frac{1}{2}\rho \left(\frac{C_D A}{m} \right) v_r \vec{v}_r$
$\vec{f}_{SRP}$	Solar radiation pressure	$\vec{f}_{SRP} = -P_{Sun} \frac{\nu A}{m} C_R \vec{u}_{SatSun}$
$\vec{f}_{ERP}$	Earth radiation pressure	$\vec{f}_{ERP} = -P_{Earth} \frac{\nu A}{m} C_R \vec{u}_{SatEarth}$
$\vec{f}_{Other}$	Other forces like thermal effects or magnetically induced forces, y-bias force	

$\vec{r}, \dot{\vec{r}}$ = satellite position and velocity vector (planeto-centric inertial frame), T_{xyz}^{XYZ} = transformation matrix between rotating axes (x, y, z) and nonrotating axis (X, Y, Z), $T_{r\phi\lambda}^{xyz}$ = transformation matrix between spherical and rectangular coordinates, U' = non-spherical gravity potential, n_p = number of perturbing bodies, $\mu_j = GM_j$ gravitational parameter of body M_j, $\vec{\Delta}$ = relative position vector of body M_j with respect to satellite position vector r_j, c = speed of light, β, γ = Post-Newtonian metric parameters, A = satellite cross sectional area, C_D = drag coefficient, m = satellite mass, $\vec{v}_r$, v_r = vector and norm of relative (satellite to atmosphere) velocity vector, P_{Sun} = momentum flux from the Sun at the satellite location, C_R = reflectivity coefficient, ν = eclipse factor (0 in Earth shadow, 1 if in sunlight), $\vec{u}_{SatSun}$ = satellite to Sun unit vector, P_{Earth} = Earth momentum flux parameter, $\vec{u}_{SatEarth}$ = satellite to Earth unit vector

with the various force contributions described in Table 6.2.

It can be readily seen that the ordinary differential equation of motion Eq. (6.2) has to be solved in order to obtain the full (6-dimensional) state vector $\vec{X}(t) = \{\vec{r}, \dot{\vec{r}}\}$ at epoch t. Analytical solutions are available only if no perturbing forces are present, i.e., $\vec{f}_p = 0$, which is referred to as the ideal two-body problem which however is not a realistic case for satellite operations.[4] To solve the equation of motion with a representative set of perturbing forces, numerical integration techniques are applied and the second order differential equation Eq. (6.2) is rewritten into a set of first order differential equations for $\vec{X}(t) = \{\vec{r}, \vec{v}\}$ by introducing the velocity vector $\vec{v}$ as an auxiliary variable, which is also known as *variation* of Cowell's formulation

---

[4] There are also special analytical solutions available for the restricted three-body problem which assumes circular orbits for the primary and secondary body and the mass of the third body (satellite) to be negligible compared to that of the major bodies (refer to e.g., [4, 5]).

$$\dot{\vec{r}} = \vec{v}$$

$$\dot{\vec{v}} = -\frac{\mu\vec{r}}{r^3} + \vec{f}_p \tag{6.4}$$

Integrating Eq. (6.4) yields the state vector $\vec{X}(t) = \{\vec{r}, \vec{v}\}$ at epoch t providing as input the initial conditions $\vec{X}_0(t) = \{\vec{r}_0, \vec{v}_0\}$. Various numerical integration techniques are available which can be categorised as follows:

- Single-step Runge-Kutta methods have the fundamental property that each integration step can be performed completely independent of one another. Such methods are therefore easy to use and the step size can be adapted each time.
- Multi-step methods reduce the total number of functions that need to be evaluated but require the storage of values from previous steps. These methods are suitable in case of complicated functions and examples are the Adams-Bashfort or Adams-Moulton methods.
- Extrapolation techniques like the Burlish-Stoer method perform one single integration step but improve the accuracy of the Runge-Kutta method by dividing this step into a number of (micro-) steps for which extrapolation methods are used in order evaluate the integration function.

Single step methods will usually be used in combination with adaptive step size techniques that perform an error estimation at each step, compare it to a predefined tolerance, and adapt the size for the subsequent step (refer to *Runge-Kutta-Fehlberg* method). Numerical integration techniques are extensively discussed in the open literature and the reader is referred to e.g., [6], [7], [8], [9], [10] to gain further insight.

6.3   Orbit Determination

The aim of this section is to present the fundamental formulation of the statistical[5] orbit determination (OD) problem and to provide an overview of the various techniques used in modern satellite operations. For a more detailed description of this complex subject matter the reader is encouraged to consult a variety of excellent text books dedicated to this specific topic (cf., [3], [11], [12], [13]).[6]

---

[5] The *statistical* orbit determination stands in contrast to the *deterministic* one which does not consider observation errors and limits the problem to only consider the required number of observations. The statistical approach in contrast processes a higher number of measurements than actually required which allows to reduce and minimise the effect of noise or errors stemming from the observations.

[6] The notation and formulation presented here follows the one given by Tapley *et. al* (cf. [3]) which should allow the reader to more easily consult this highly specialised text.

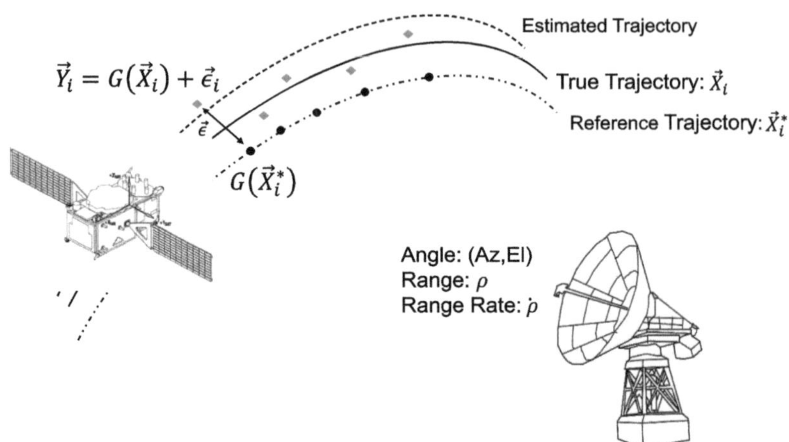

Fig. 6.2 Basic concept of Statistical Orbit Determination: measured observations are represented by the grey diamonds, modelled observations by black circles

The *true trajectory* of a satellite flying in an orbit around a central body is only known to a certain degree of accuracy at any given time. It is therefore common to term the current best knowledge trajectory as the *reference trajectory* which is also the starting point of any OD process. The objective of OD is to improve the accuracy of the reference trajectory and to minimise its deviation from the true one by processing new available observations (see Fig. 6.2). Before outlining the basic OD formulation, it is worth to define a few relevant terms and their notation in more detail.

- The n-dimensional vector $\vec{X}(t) \equiv \vec{X}_t = \{\vec{r}(t), \vec{v}(t)\}$ contains the position and velocity vectors of the satellite at epoch t. To refer to a state vector at a different epoch t_i, the short notation $\vec{X}(t_i) \equiv \vec{X}_i$ is used.
- The observation modelling function $\vec{G}(\vec{X}_i)$ defines the mathematical expression to compute the observation at epoch t_i, with $(i = 1, \ldots, l)$, using the available state vector information $\vec{X}_i$. The asterix (e.g., $\vec{X}_i^*$) indicates that the state vector stems from the reference trajectory at epoch t_i.
- The observation vector $\vec{Y}_i$ represents the measured observations at epoch t_i which can be used as input to the OD process to update the orbit knowledge. The dimension $i = 1, .., l$ represents the number of different epochs at which observations are available and the dimension of $\vec{Y}_i$ (i.e., $1, .., p$) is determined by the number of different types of observations available. As an example, if azimuth, elevation, range, and range rate measurements would be provided by the TT&C station, then $p = 4$. In general, $p < n$, where n is the number of parameters or unknowns to be determined, and $p \times l >> n$.
- The residual vector $\vec{\epsilon}_i$ has also dimension p and is given at epochs t_i (again with $i = 1, .., l$) and contains the differences between the measured observations and the modelled ones. The measured values are affected by measurement

errors of the tracking station whereas the computed ones by inaccuracies of the observation modelling function and the deviation of the reference trajectory from the true one.

With this notation in mind, the governing relations of the OD problem can be formulated as

$$\dot{\vec{X}}_i = F(\vec{X}_i, t_i)$$
$$\vec{Y}_i = G(\vec{X}_i, t_i) + \vec{\epsilon}_i \tag{6.5}$$

where the first equation has already been introduced in Eq. (6.4). It is important to note that both the dynamics $\dot{\vec{X}}$ and the measurements $\vec{Y}$ involve significant nonlinear expressions introduced by the functions F and G. This makes the orbit determination of satellite trajectories a nonlinear estimation problem that requires the application of nonlinear estimation theory. With the assumption that the reference trajectory $\vec{X}_i^*$ is close enough to the true trajectory $\vec{X}_i$ throughout the time interval of interest, Eq. (6.5) can be expanded into a Taylor's series about the reference trajectory. This leads to the following two expressions:

$$\dot{\vec{X}}(t) = F(\vec{X}, t) = F(\vec{X}^*, t) + \left[\frac{\partial F(t)}{\partial \vec{X}(t)}\right]^* [\vec{X}(t) - \vec{X}^*(t)]$$
$$+ O_F[\vec{X}(t) - \vec{X}^*(t)] \tag{6.6}$$

$$\vec{Y}_i = G(\vec{X}_i, t) + \vec{\epsilon}_i = G(\vec{X}_i, t_i) + \left[\frac{\partial G}{\partial \vec{X}}\right]_i^* [\vec{X}(t_i) - \vec{X}^*(t_i)]$$
$$+ O_G[\vec{X}(t_i) - \vec{X}^*(t_i)] + \epsilon_i \tag{6.7}$$

where the symbols $O_F(\ldots)$ and $O_G(\ldots)$ represent the higher order terms of the Taylor series which are neglected in the further treatment as they are assumed to be negligible compared to the first order terms.

The *state deviation vector* $\vec{x}_i$ (dimension $n \times 1$) and the *observation deviation vector* $\vec{y}_i$ dimension ($p \times 1$) can now be introduced as

$$\vec{x}_i \equiv \vec{x}(t_i) = \vec{X}(t_i) - \vec{X}^*(t_i)$$
$$\vec{y}_i \equiv \vec{y}(t_i) = \vec{Y}(t_i) - \vec{Y}^*(t_i) = \vec{Y}(t_i) - G(\vec{X}_i^*, t_i) \tag{6.8}$$

with subscript $i = 1, .., l$ representing the different times or epochs at which observations are available.

Introducing the mapping matrices

$$
A_i^{[n,n]} = \left[\frac{\partial F_n}{\partial \vec{X}_n} \right]^* (t_i)
$$

$$
\tilde{H}_i^{[p,n]} = \left[\frac{\partial G_p}{\partial X_n} \right]^* (t_i)
$$

(6.9)

and using the deviation vectors from Eq. (6.8), the expressions in Eq. (6.6) and Eq. (6.7) can now be re-written as a set of linear differential equations

$$
\dot{\vec{x}}_i = A_i \, \vec{x}_i
$$

$$
\vec{y}_i = \tilde{H}_i \, \vec{x}_i + \vec{\epsilon}_i
$$

(6.10)

It should be kept in mind that the matrix $\tilde{H}_i$ is a $p \times n$ dimensional matrix at epoch t_i where the number of rows $(1, .., p)$ reflect the number of different types of observations available and the number of columns $(1, .., n)$ the corresponding partial derivatives with respect to the state vector components. As there will be observations available at a set of different observation epochs $i = 1, \ldots, l$, the fully composed $\tilde{H}$ matrix is further expanded to dimension $[p \times l, n]$.

$$
\tilde{H}^{[p \times l, n]} = \left\{ \begin{matrix} \tilde{H}_{t_1}^{[p \times n]} \\ \cdots \\ \tilde{H}_{t_l}^{[p \times n]} \end{matrix} \right\}
$$

(6.11)

A general solution for Eq. (6.10) can be expressed as

$$
\vec{x}(t) = \Phi(t, t_k) \, \vec{x}_k
$$

(6.12)

where $\Phi(t, t_k) = \frac{\partial \vec{X}(t)}{\partial \vec{X}(t_k)}$ is the $n \times n$ *state transition matrix*, which can be computed in the same numerical integration process as the state vector solving the following set of differential equations, known as *variational equations*:

$$
\dot{\Phi}(t, t_k) = A(t) \, \Phi(t, t_k)
$$

(6.13)

with the initial conditions $\Phi(t_k, t_k) = I$, I being the identity matrix.

The state transition matrix allows to relate the observation deviation vector $\vec{y}_i$ to the state vector $\vec{x}_k$ at any single epoch t_k

$$
\vec{y}_i = \tilde{H}_i \, \Phi(t_i, t_k) \, \vec{x}_k + \vec{\epsilon}_i
$$

(6.14)

With the definition of $H = \tilde{H} \, \Phi$ and dropping the subscripts, the above equation can be simplified to

$$\vec{y} = H\,\vec{x} + \vec{\epsilon} \tag{6.15}$$

The objective of the orbit determination process is to find the state deviation vector $\hat{\vec{x}}$ that minimises the observation deviation vector $\vec{y}$, or more precisely, the sum of the squares of $(\vec{y} - H\,\vec{x})$. The linearisation of the problem in the form of Eq. (6.15) allows the application of linear estimation theory algorithms like the weighted least squares method which provides

$$\boxed{\hat{\vec{x}}_k = (H^T\,W\,H)^{-1}\,H^T\,W\,\vec{y} = P_k\,H^T\,W\,\vec{y}} \tag{6.16}$$

where W is the so called weighting matrix that contains as diagonal elements w_i the normalised weighting factors (range between zero and one) for each of the observation vectors. $P_k \equiv (H^T\,W\,H)^{-1}$ is the $n \times n$ *covariance matrix* providing an estimate on the accuracy of the estimation vector $\hat{\vec{x}}$. In other words, the larger the magnitude of the elements of the matrix P_k, the less accurate is the estimate.

It is worth to mention that orbit determination is not a one time activity during daily satellite operations but must be repeated whenever a new set of observations (radiometric data) is received from the TT&C stations. It is therefore important to implement a mathematical formulation that is able to consider an *a priori* state and the corresponding covariance matrix from a previous run. Such a formulation is referred to as *minimum variance method* and a more detailed treatment is available in the cited literature (e.g., [3]). In the same context it is also relevant to consider OD formulations that either ingest a set of observations in one run (referred to as *batch processors*) or sequentially, whenever they become available (referred to as *sequential estimation algorithm* or *Kalman filter*).

6.4 Orbit Control

Orbit control encompasses the regular monitoring of the current orbit evolution, the detection of the need for an orbit correction, and the planning and execution of orbit manoeuvres. The main phases of the orbit monitoring and control process are schematically shown in Fig. 6.3. The starting point for every orbit control activity should always be the acquisition of the up-to-date satellite orbit from an orbit determination that is based on the most recent available radiometric data. With the up-to-date orbit knowledge in place, the orbit propagation and monitoring process can be performed by comparing the current Keplerian elements with the target ones. If the deviation is considered to be too large and the satellite is reaching the borders of its station keeping control box, the manoeuvre planning activity can start to determine the required magnitude of orbit change which can be expressed as differential Keplerian elements, Δa, Δe, Δi, $\Delta\Omega$, and Δu. For circular orbits (near-zero eccentricity), the Gauss equations can be used to derive the size, direction, and location of the orbit change manoeuvre Δv

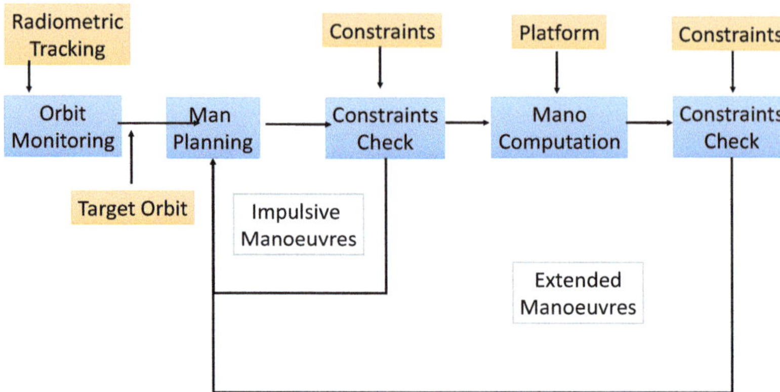

Fig. 6.3 Orbit monitoring and correction flow chart including constraints checks (refer to text)

$$
\begin{aligned}
\Delta a &= \frac{2a}{v}\Delta v_T \\[2mm]
\Delta e_x &= 2\cos u\,\frac{\Delta v_T}{v} + \sin u\,\frac{\Delta v_R}{v} \\[2mm]
\Delta e_y &= 2\sin u\,\frac{\Delta v_T}{v} - \cos u\,\frac{\Delta v_R}{v} \\[2mm]
\Delta i &= \cos u\,\frac{\Delta v_N}{v} \\[2mm]
\Delta\Omega &= \frac{\sin u}{\sin i}\,\frac{\Delta v_N}{v}
\end{aligned}
\tag{6.17}
$$

where quantities a, i, and Ω refer to the classical Keplerian elements semi-major axis, inclination, and the right ascension of the ascending node, the Δv subscripts R, T, and N refer to respectively the radial, tangential, and normal direction in the orbital (LVLH) reference frame (see Appendix A.3), and the eccentricity vector $\vec{e} = \{e_x, e_y\}$ is defined by

$$
\begin{aligned}
e_x &= e\cos(\omega) \\
e_y &= e\sin(\omega)
\end{aligned}
\tag{6.18}
$$

and points from the orbit apoapsis to its periapsis with its magnitude providing the value of the orbit eccentricity. The argument of latitude (also referred to as phase)

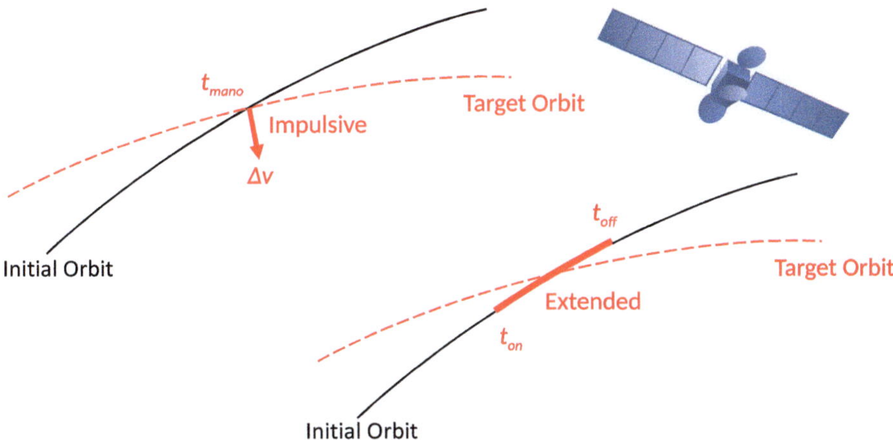

Fig. 6.4 Impulsive and extended manoeuvre definitions

is given by $u = \omega + v$ and can be readily obtained from the vector separation angle between the satellite position $\vec{r}$ and the ascending node vector $\vec{n}$.[7]

The Gauss equations as defined in Eq. (6.18) provide orbit corrections as so called *impulsive manoeuvres* which can be understood as instantaneous velocity changes in Δv of a given magnitude, direction (e.g., in-plane or out-of plane), and time or orbit location as depicted in the upper portion of Fig. 6.4. The important characteristic of such impulsive manoeuvres is that they are fully determined by the extend of required orbit change. They are however fully independent on satellite specific parameters like mass or thrust force and are therefore referred to as *platform independent manoeuvres*. To give an (extreme) example: to change the semi-major axis for a given orbit geometry by a certain extent, the exact same Δv value applies for a nano-satellite with a mass of a few kg and the International Space Station with a mass of several metric tones.

Once a first impulsive manoeuvre strategy has been determined, it needs to be checked against any known satellite specific constraints for manoeuvre execution which is indicated by the "Constraint Check" box of Fig. 6.4. Examples of constraints could be the visibility of Sun or Moon in AOCS sensors, eclipse periods affecting the power budget, Solar incidence angles imposed by thermal load limits, or required visibility to one or more TT&C stations. In case of a violation the manoeuvre strategy needs to be revised which is indicated by the loop over "Impulsive Manoeuvres".

After the impulsive manoeuvre strategy has converged, the platform specific *extended manoeuvre* parameters can be derived, which is indicated by the "Mano

---

[7] The argument of latitude is the preferred quantity for the definition of the satellite position along the orbit as it is also suitable for near-circular orbits for which the perigee is not well defined.

Computation" (see Fig. 6.4). The extended manoeuvre parameters are the ones required for the actual commanding and usually comprise values for the start epoch, thruster burn duration (or thruster on and off times), and the required satellite attitude during burn.[8] For larger sized manoeuvres the burn duration might last up to several minutes and an additional constraint check is highly recommended to ensure that the constraints are fulfilled for the full burn duration. As an example, the start epoch of a manoeuvre could be well outside of an eclipse, but the satellite might enter into an eclipse period during the burn which could violate a flight rule and make the manoeuvre placement invalid. Only if all constraint checks are fulfilled with extended manoeuvres, the manoeuvre command parameters can be computed and uplinked to the satellite.

The burn duration Δt_{thrust} of the extended manoeuvre can be derived from the Δv value using Newton's second law

$$
\boxed{\Delta t_{thrust} = \frac{m_{sat}\,\Delta v}{n_t\,\bar{F}\,\cos(\alpha)\,\eta_{eff}}}
\tag{6.19}
$$

where $\bar{F}$ is the average thrust force of one single satellite thruster, n_t is the number of thrusters used for the manoeuvre burn, α is the thruster tilt angle, and η_{eff} is an efficiency factor to compensate for (known or expected) manoeuvre execution errors. The number of thrusters used for an orbit correction manoeuvre depends on the satellite AOCS software implementation and usually fewer thruster are used for small sized manoeuvres.

The thrust force F and the mass flow $\dot{m}$ depend on the tank pressure which is a function of the filling ratio. The exact relationship for $F(p)$ and $\dot{m}_f(p)$[9] are hardware specific and the exact dependencies can only be measured on ground in a dedicated thruster performance measurement campaign. This data is therefore an important delivery from the space to the ground segment[10] and needs to be implemented in the FDF in order to ensure an accurate computation of the manoeuvre command parameters.

The satellite mass m_{sat} decreases during each manoeuvre burn due to the consumption of propellant which has to be taken into consideration when applying Eq. (6.19). The decrease of propellant mass m_p can be estimated using the well known rocket or *Tsiolkovski* equation.

---

[8] The detailed list of the required manoeuvre command parameters need to be consulted from the applicable satellite manuals.

[9] The mass flow rate is needed to compute the specific impulse: $I_{sp} = \frac{F\,\cos\alpha}{\dot{m}_f\,g}$.

[10] This is referred to as the *thruster performance model* and contractually handled as a *Customer Furnished Item* or CFI.

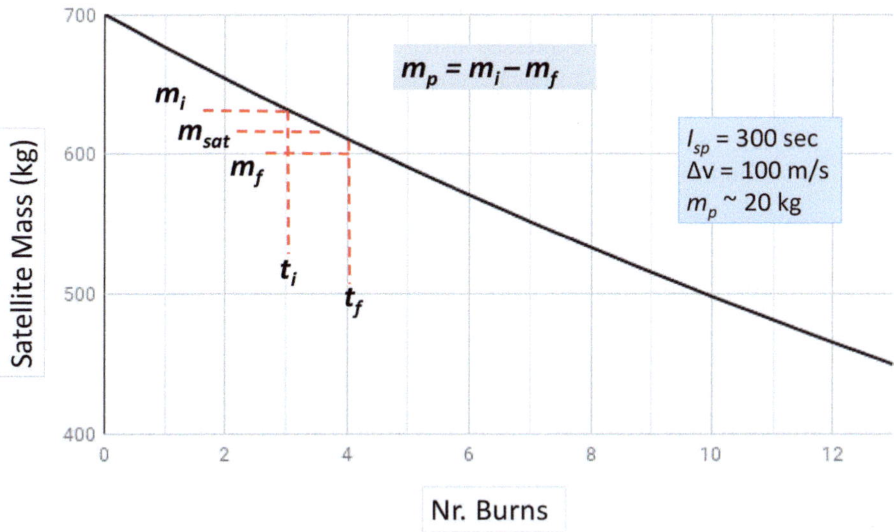

Fig. 6.5 Numerical application of the rocket equation Eq. (6.20) for a series of equally sized burns of Δv=20 m/sec; m_i and m_f refer to the initial and final satellite mass and m_p is the consumed propellant mass at each burn

$$m_p = m_i - m_f = m_i \left[1 - e^{-\frac{\Delta v}{I_{sp}\, g}} \right] \tag{6.20}$$

where m_i and m_f are the satellite mass at the beginning and the end of the burn, I_{sp} is the specific impulse of the thruster system (propellant specific), and g is the gravitational constant. A numerical application of the rocket equation is demonstrated in Fig. 6.5 where the decrease in satellite mass for a series of (equally sized) manoeuvres with a Δv value of 100 m/s is shown. It can be seen that even for a considerable sized Δv, a linear interpolation of the mass decrease curve is an acceptable approximation in order to estimate the satellite mass at the manoeuvre mid-point. m_{sat} in Eq. (6.19) can therefore be approximated with the mid-point satellite mass

$$m_{sat} = m_i - \frac{m_p}{2} = \frac{m_i}{2} \left[1 + e^{-\frac{\Delta v}{I_{sp}\, g}} \right] \tag{6.21}$$

The required satellite attitude during a manoeuvre burn phase depends on the specific orbit element that needs to be changed. From Eq. (6.17) it can be readily seen that a change of the orbit semi-major axis, eccentricity, or satellite phase requires an in-plane manoeuvre with the thrust directed parallel to the orbital frame's tangential direction $\vec{e}_T$. A change of inclination or ascending node in contrast requires an out-of-plane manoeuvre with a thrust vector directed parallel to the orbital frame normal direction $\vec{e}_N$.

The execution time of a burn is derived from the required location of the burn on the orbit which can be expressed using the argument of latitude u. The last two equations of Eq. (6.17) indicate that a pure inclination change requires the manoeuvre to be located in the ascending node (i.e., $u = 0$ deg), whereas a pure change of Ω requires a manoeuvre location at the pole (i.e., $u = 90$ deg). It is also possible to combine the change of both elements using the following relation

$$\Delta v_N = \sqrt{(\Delta i)^2 + (\Delta \Omega \, \sin i)^2}$$
$$u = \arctan\left(\frac{\Delta \Omega \, \sin i}{\Delta i}\right) \tag{6.22}$$

Despite the accurate calculation and commanding of an orbit correction manoeuvre, its actual execution in orbit might not lead to the expected orbit change. This is mainly due to the *manoeuvre execution error* caused by the satellite. Especially for the first manoeuvres this error is more significant and can have a size of up to 5% of the commanded Δv value and be attributed to a different behaviour of the satellite's AOCS system in orbit compared to what has been modelled on ground. It is therefore important to use the first post-manoeuvre TT&C tracking data to estimate that execution error. This can be done with an orbit determination that includes the commanded manoeuvre in the orbit modelling equations which then allows to estimate the actually achieved one together with the orbit. This method is referred to as *manoeuvre calibration* and is a very important means to characterise the in orbit thruster performance which can then be expressed in an efficiency factor η_{eff} (see Eq. 6.19) that can be considered for subsequent manoeuvre planning.

Orbit correction manoeuvres are required in almost all orbit phases and the following terminology is widely used that is mainly based on their size and purpose:

- *Orbit acquisition* manoeuvres are performed during a LEOP phase and executed shortly after separation of the satellite from the launcher's upper stage. The targeted orbit changes, both in-plane and out-of-plane, are usually quite significant. The main reason for this is that a launchers usually place a satellite into an *injection orbit* that is by intention different from the operational orbit in order to avoid any placement of harmful objects (e.g., the launcher's upper stage or dispensers) into the same region, which could later pose a collision risk to the satellite.
- *Station-keeping* manoeuvres are usually small size manoeuvres meant to keep the satellite within a certain orbit control box. Such manoeuvres play an important role for satellites placed in geostationary orbit or constellations with clearly defined geometrical patterns (e.g., Walker constellations [14]).
- *Disposal* manoeuvres are planned at the end-of-life of a satellite with the aim to place the satellite into a graveyard orbit which (so far) is not considered as relevant for standard satellite services. For satellites in low altitude orbits, the aim is to further reduce the semi-major axis in order to accelerate their reentry and make them burn up in the upper layers of the atmosphere.

- *Fine-positioning* or *clean-up* manoeuvres are planned and executed in order to correct manoeuvre execution errors and perform very accurate orbit positioning.

6.5 Propellant Gauging

A satellite's operational lifetime is strongly dependant on the amount of remaining propellant available for future orbit and attitude control. Methods for estimating the current (and remaining) fuel budget are therefore of great importance, especially for service oriented satellite projects like TV broadcasting, telecommunication, or navigation, for which every year of service provision relates to significant commercial revenues. The knowledge of the current propellant mass is also technically relevant as it influences the overall satellite mass and its distribution. This impacts the computation of the thruster burn duration as defined in Eq. (6.19), the location of the centre of mass (CoM), and the satellite's inertia matrix. Methods for fuel estimations are frequently referred to as *propellant gauging methods* and need to be implemented in every ground segment. The two most commonly used algorithms are the *Book-Keeping* and the *Pressure-Volume-Temperature* (PVT) method, both are briefly described below.

6.5.1 Book-Keeping Method

This method requires the knowledge of the thruster on-times $t_{on,i}$ of each thruster cycle (activation) which should be available from satellite housekeeping telemetry. The consumed propellant mass of one active thruster cycle can be computed from

$$m_{p,i} = \dot{m}_i \cdot t_{on,i} \cdot \eta_{\dot{m},i} \tag{6.23}$$

where $\dot{m}_i$ is the mass flow rate (kg/sec) and $\eta_{\dot{m},i}$ is the dimensionless mass flow efficiency factor which considers a change of the mass flow value if thrusters are operated in pulsed mode. The consumed propellant at any time t during the mission is given by summing up the consumed propellant mass of all thruster cycles n that have so far occurred

$$m_p(t) = \sum_{i=1}^{n} m_{p,i} \tag{6.24}$$

The remaining propellant mass m_p can now be derived from the initial propellant mass at Beginning-of-Life (BOL) $m_{p,\,BOL}$ which is measured when the satellite is being fuelled on ground

$$m_p = m_{p,BOL} - m_p(t) \qquad (6.25)$$

Eq. (6.25) makes the book keeping aspect of this technique quite obvious but it also shows its major weakness of accumulating the prediction error with every summation step. This implies the lowest fuel estimation accuracy at the satellite's end of life, when the highest accuracy would actually be needed.

6.5.2 PVT Method

The detailed mathematical formulation of the PVT method depends on the satellite platform specific tank configuration, but the main underlying principle is based on the tank pressure and temperature measured by dedicated transducers and provided as part of the housekeeping telemetry. The additional knowledge of the tank volume from ground acceptance tests and the use of the ideal gas law allows to derive the propellant mass. A simplified formulation for a typical mono-propellant blow-down propulsion system is presented here for which the notation from NASA-TP-2014-218083 [15] has been adapted. A schematic drawing that helps to understand the notation and method is provided in Fig. 6.6.

The supply bottle on the right side contains a pressurant gas for which Helium is frequently used having the important property to be non-condensable with the liquid propellant. The supply bottle is connected to the tank that contains the liquid propellant. When the latch valve is opened, the gaseous He can flow into the ullage and initiates the flow of the liquid propellant to the catalyst bed of the thruster where it ignites. The objective of the propellant gauging algorithm is to compute the mass of the liquid propellant

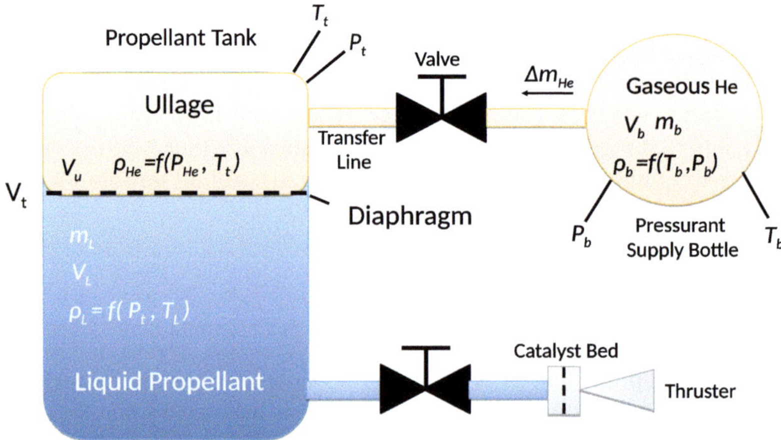

Fig. 6.6 Main components of a blow-down propulsion system

$$m_L = \rho_L \, V_L = \rho_L \, (V_t - V_u) \tag{6.26}$$

where V_t is the propellant tank volume that needs to be accurately measured prior to launch and V_L and V_u are the volumes of the liquid propellant and the ullage sections respectively. V_u increases with the mass flow of gaseous He from the pressurant bottle into the propellant tank Δ_m according to

$$V_u = \frac{\Delta m_{He}}{\rho_{He}} \tag{6.27}$$

where the He density ρ_{He} is a known function of the measured tank temperature T_t and the partial He pressure P_{He} inside the ullage section, i.e.,

$$\rho_{He} = f\,(P_{He}, T_t) \tag{6.28}$$

P_{He} can be derived from the measured tank pressure P_t and the known value of the propellant's vapour saturation pressure P_{sat} according to

$$P_{He} = f\,(P_t - P_{sat}, T_t) \tag{6.29}$$

The amount of gaseous He mass flown into the ullage can be expressed as

$$\Delta m_{He} = m_{He,i} - m_{He,t} = V_b(\rho_{He,i} - \rho_{He,t}) \tag{6.30}$$

where $m_{He,i}$ is the Helium mass at initial state and $m_{He,t}$ the one at time t when the PVT method is being applied. The volume of the supply bottle V_b is a known quantity and the gas densities at initial time and time t are a function of the measured pressure and temperature values at these times, i.e.,

$$\begin{aligned} \rho_{He,i} &= f(T_i, P_i) \\ \rho_{He,t} &= f(T_b, P_b) \end{aligned} \tag{6.31}$$

The accuracy of the PVT method depends on various factors like the uncertainty in the initial loading condition, pressurant gas solubility in propellant,[11] thermal condition that influence the stretch of the tank volume, and finally the measurement accuracy of the pressure and temperature transducers.

---

[11] To avoid such an unwanted interaction, some tanks implement a diaphragm that separates the pressurant gas from the propellant in order to avoid such interaction.

6.5.3   *Gauging Accuracy*

The accuracy of a gauging method is an important factor when deciding which method to use to predict the satellite EOL. Several studies on this topic have been published and the reader his encouraged to consult them in more detail (cf., [16], [17], [18]). A generic conclusion is that the accuracy of both the PVT and the book keeping method are quite reliable at the beginning of life but decrease with time.

An entirely different approach is based on the thermal response (capacitance) of the propellant tank to heating and the comparison of the measured values to simulated ones from a tank thermal model [19]. This method shows the opposite behaviour with an increase in accuracy with decreasing propellant load and can therefore be considered as a complementary method to be applied at the EOL.

6.6   Onboard-Orbit-Propagator

The onboard software requires the knowledge of the satellite's state vector in order to derive frame transition matrices and to predict various orbit events like upcoming eclipses (see also Chap. 3). Satellites therefore implement their own orbit propagation module which is referred to as *onboard orbit propagator* or OOP. Such algorithms either implement orbit propagation techniques based on numerical integration with representative force model as described in Sect. 6.2 or use more simplistic approaches that limit themselves to interpolation techniques. The orbit propagation performed by the satellite will by design be of much lower accuracy compared to the one performed on ground, simply to reduce the computational load on the onboard processor and its memory. This also implies that the accuracy of the satellite generated orbit solution will decrease more rapidly and sooner or later reach a point when it is not reliable anymore. At this point, or preferably even earlier, it requires to receive updated orbit elements from ground so it can restart the propagation with new and accurate initial conditions. The satellite manufacturer has to provide a dedicated telecommand with an adequate description of the OOP design, type of update (e.g., Kepler elements, cartesian state vector, reference frame, etc.) and required update frequency as part of satellite user manual.

Despite the higher accuracy of the orbit propagation performed on ground, it is still recommended to implement an OOP emulator in addition. This allows to simulate the orbit information available to the satellite onboard software which might be relevant for the planning of specific activities that depend on event predictions times determined by the satellite.

6.7 Collision Monitoring

Due to the growing complexity of the orbital environment, space assets face an increasing risk to collide with space debris, other satellites (retired and active ones), or the upper stage of the own launch vehicle after separation. The ability to perform collision risk monitoring has therefore become a vital task for a modern ground segment in order to guarantee the safety of a satellite project.

As the collision monitoring module requires numerical orbit propagation techniques, it is usually integrated in the FDF where orbital mechanics libraries can be easily accessed. A conceptual view of a collision monitoring module is depicted in Fig. 6.7 where all the input data are shown on the left side and comprise the following files

- A space object catalogue providing a database of all tracked space objects usually in the form of Two-Line-Element (TLE). The contents of this database needs to be provided by external sources like NORAD (provided via the Center for Space Standards & Innovation [21]) or the EU Space Surveillance and Tracking (SST) framework [22] who keep an up-to-date database of the highly dynamic space environment. As TLEs by design have only a limited validity of a few days, a regular update mechanism for this catalogue is necessary in order to guarantee a complete and up-to-date scanning of all space objects.
- An accurate estimate of the satellite's orbit data for which collision monitoring is performed. This should be based on the most recent orbit determination which not only provides the state estimate but also its covariance matrix. The latter one can be seen as an uncertainty ellipsoide that is centered on the satellite position estimate.

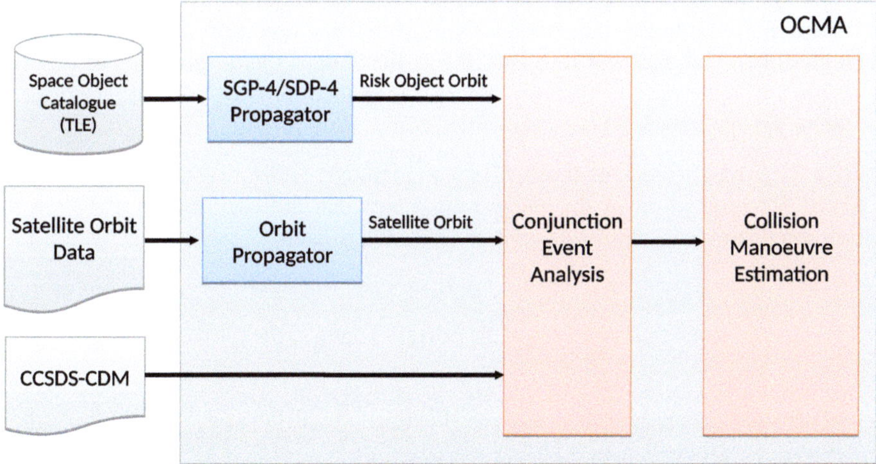

Fig. 6.7 Schematic overview of collision monitoring functionality in FDF. CCSDS CDM = Collision Data Message as defined by CCSDS [20]

- The CCSDS *Conjuction Data Message* or CDM [20] is a specific format that has been defined to allow an easier exchange of spacecraft conjunction information between conjunction assessments centres and satellite owners or operators. This standardised message contains all relevant information for both objects involved in a conjunction at the *time of closest approach* (TCA). This comprises state vectors, covariance matrices, relative position, velocity vectors, and other relevant information that is useful for further analysis by the recipient. It is highly recommended to implement such an external interface and sign up with one of the centres providing this type of messages (e.g., the Joint Space Operations Center, ESA/ESOC Space Debris Office, or EUSST).

The SGP-4/SDP-4[12] propagator component implements the necessary algorithms that allow the orbit propagation of the TLE based *risk* or *chaser* object to the required epoch. The Conjunction Event Analysis component in Fig. 6.7 contains the algorithms to determine a possible conjunction of the satellite and the risk object and also determines the likelihood of a collision which is achieved via two principle methods. The first one is the successive application of the following set of filters (cf. [23] and [24]):

1. Unreliable TLEs: a filter that rejects TLEs with epochs that are too old to be still be reliable (e.g., older than 30 days w.r.t. the current investigation epoch) or have decayed already.
2. Altitude filter: rejects TLEs for which the risk object's orbit does not intersects the spherical altitude shell between the perigee and apogee of the target orbit (also considering the orbit decay and short-periodic semi-major axis variations).
3. Orbit geometry check: computes the intersection line of the risk and target orbits and rejects those TLEs whose closest approach distance at the ascending and descending nodes is larger than the maximum extension of the user defined collision reference ellipsoid that is centred on the target object.
4. Phase filter: computes the passage times of risk and target object at the ascending nodes of the intersection line and rejects the TLE if the at the times of possible conjunction the relative distance is larger than the reference collision ellipsoid of the target spacecraft.

If a TLE has passed all these filters an iterative Newton scheme is applied to find the zero-transition of the range-rate $\dot{\rho}$ between the two objects at which the minimum distance is achieved (time of closest approach)

$$\dot{\rho} = \Delta\vec{v}\frac{\Delta\vec{r}}{\Delta r} = 0.0 \quad \rightarrow t_{TCA} \tag{6.32}$$

The second method requires less CPU time as it replaces the last two filters with the implementation of a smart sieve algorithm (cf., [25] and [26]) which analyses

---

[12] The SDP-4 theory is used for objects with an orbital period exceeding 225 minutes.

the time history of ranges $\vec{\rho}(t)$ between the target orbit $\vec{\rho}_t$ and the risk orbit $\vec{\rho}_r$ at equidistant time steps Δt across the prediction interval. At each time step the components of the range vector are sequentially checked against a series of safety distances, that are refined each time (refer to e.g., Table 8.2 of [27]). For the orbits passing the smart sieve filter, the root-finding method defined in Eq. (6.32) is applied again to get the time of closest approach t_{TCA}. To assess the collision probability, the covariance matrices of both conjunction objects are propagated to the conjunction time t_{TCA}. Reducing the (7 x 7) state covariance matrix to a (3 x 3) position covariance matrix and assuming that these are not correlated, they can be simply added to obtain a combined covariance matrix [28].

$$\grave{C} = \grave{C}_t(t_{TCA}) + \grave{C}_r(t_{TCA}) \tag{6.33}$$

This 3D combined covariance matrix is then projected onto the 2D B-plane, which is centred at the target body position and perpendicular to the relative velocity at $\Delta \vec{v}_{TCA}$. The collision probability can then be obtained via the surface integral of the probability density function.[13]

6.8 Interfaces

The Flight Dynamics Facility needs to maintain a large number of interfaces to different elements of the GCS but also to external entities like the satellite manufacturer. A generic overview is provided in Fig. 6.8 and the various signal flows detailed below.

- **SATMAN** (see Chap. 3): an interface between the FDF and the satellite manufacturer is not mandatory but useful for projects that need to deploy a large number of satellites as a constellation. The purpose of such an interface is to exchange satellite specific configuration data like mass (dry and wet), center of mass, or inertia matrices. An automatic ingestion and processing functionality at FDF level would also avoid the need of any manual operator interaction reducing the risk of incorrect data transfer.
- **TT&C** (see Chap. 5): the interface to the TT&C is needed for the provision of pointing files to the antennas and the reception of radiometric, angle and meteorological data in return.
- **SCF** (see Chap. 7): the FDF sends to the SCF any flight dynamics specific command parameters. Main candidates are relevant inputs for the OOP update, orbit manoeuvre correction parameters, or sensor inhibition parameters. The format will very much depend on the software deployed on the SCF system (for SCOS based systems the Task Parameter File (TPF) format is usually required). In return the SCF provides to the FDF any flight dynamics relevant telemetry like

---

[13] For a more detailed mathematical explanation the reader is referred to [27] and [28].

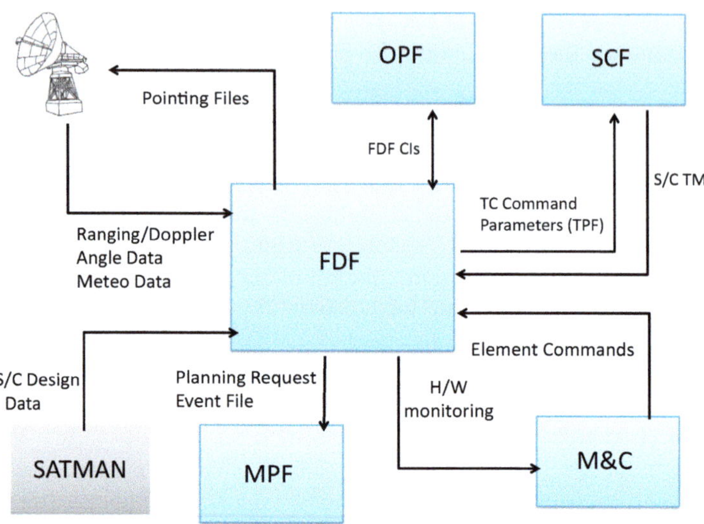

Fig. 6.8 Overview of FDF specific interfaces

e.g., thruster on-times required for the computation of the mass consumption via the Book Keeping method (see Sect. 6.5).

- **MPF** (see Chap. 8): the FDF needs to provide an event file to the Mission Planning Facility that contains station visibilities and the timings of any other orbit specific events. Furthermore, the FDF needs to be able to inform the MPF about flight dynamics relevant planning requests (e.g., OOP update, or manoeuvres).
- **OPF** (see Chap. 9): as the FDF maintains a significant database of flight dynamics specific parameters, an interface to the OPF helps to ensure the proper configuration control of this data. The data to be exchanged are in the form of Configuration Items (CIs) that need to be clearly defined as part of that specific interface.
- **MCF** (see Chap. 10): the interface to the Monitoring and Control Facility serves to monitor the state of the FDF and allows the transmission of macros for remote operations (e.g., start of automatic procedures during routine operations phases).

References

1. European Space Agency. (2017). *SCOS-2000 Database Import ICD*. EGOS-MCS-S2k-ICD-0001, v7.0.
2. International Earth Rotation Service. (2022). https://www.iers.org/IERS/EN/Home/home_node.html. Accessed 09 March 2022.
3. Tapley, B. D. (2004). *Statistical orbit determination*. Cambridge: Elsevier Academic Press.

4. Szebehely, V. (1967). *Theory of orbits: The restricted problem of three bodies*. New York and London: Academic.

5. Vallado, D. A. (2004). *Fundamentals of astrodynamics and applications, space technology library* (STL) (2nd ed.). Boston: Microcosm Press and Kluwer Academic Publishers.

6. Shampine, L. F., & Gordon, M. (1975). *Computer solution of ordinary differential equations: The initial value problem.* San Francisco, CA: Freeman.

7. Bulirsch, R., & Stoer, J. (1966). Numerical treatment of ordinary differential equations by extrapolation methods. *Numerical Mathematics, 8,* 1–13.

8. Hairer, E., Norsett, S. P., & Wanner, G. (1987). Solving ordinary differential equations. Berlin, Heidelberg, New York: Springer.

9. Gupta, G. K., Sacks-Davis, R., & Tischer, P. E. (1985). A review of recent developments in solving ODEs. *Computing Surveys, 17,* 5.

10. Kinoshita, H., & Nakai, H. (1990). Numerical integration methods in dynamical astronomy. *Celestial Mechanics, 45,* pp. 231–244.

11. Montenbruck, O., & Gill, E. (2011). *Satellite orbits: Models, methods and applications.* New York: Springer.

12. Wiesel, W. E. (2010). Modern orbit determination (2n ed.). Scotts Valley: CreateSpace.

13. Milani, A., & Gronchi, G. (2010). *Theory of orbit determination.* Cambridge: Cambridge University Press.

14. Walker, J. G. (1984). Satellite constellations. *Journal of the British Interplanetary Society, 37,* 559–571.

15. Neil, T., Dresar, V., Gregory, A., et al. (2014). *Pressure-volume-temperature (PVT) gauging of an isothermal cryogenic propellant tank pressurized With gaseous helium.* NASA/TP—2014-218083.

16. Lal, A., & Raghunnandan, B. N. (2005). Uncertainty analysis of propellant gauging systems for spacecraft. *Journal of Spacecraft and Rockets, 42,* 943–946.

17. Hufenbach, B., Brandt, R., André, G., et al. (1997). Comparative assessment of gauging systems and description of a liquid level gauging concept for a spin stabilised spacecraft. In *European Space Agency (ESTEC), Procedures of the Second European Spacecraft Propulsion Conference,* 27–29 May, ESA SP-398.

18. Yendler, B. (2006). Review of propellant gauging methods. In *44th AIAA Aerospace Sciences Meeting and Exhibit,* 9–12 Jan 2006.

19. Dandaleix, L., et al. (2004). Flight validation of the thermal propellant gauging method used at EADS astrium. In *Proceedings of the 4th International Spacecraft Propulsion Conference,* ESA SP-555, 2–9 June 2004.

20. The Consultative Committee for Space Data Systems. (2018). *Conjunction data message.* CCSDS 508.0-B-1, Blue Book (including Technical Corrigendum June 2018).

21. Center for Space Standards & Innovation. (2022). http://www.celestrak.com/NORAD/ elements/ Accessed 09 March 2022.

22. EU Space Surveillance & Tracking Framework. (2022). https://www.eusst.eu/. Accessed 09 March 2022.

23. Hoots, F., Crawford, L., & Roehrich, R. (1984). An analytical method to determine future close approaches between satellites. *Celestial Mechanics, 33,* 8.

24. Klinkrad, H. (1997). One year of conjunction events of ERS-1 and ERS-2 with objects of the USSPACECOM catalog. In *Proceedings of the Second European Conference on Space Debris* (pp. 601–611). ESA SP-393.

25. Alarcón, J. (2002). *Development of a Collision Risk Assessment Tool.* Technical Report, ESA contract 14801/00/D/HK, GMV.

26. Escovar Antón, D., Pérez, C. A., et al. (2009). closeap: GMV's solution for collision risk assessment. In *Proceedings of the Fifth European Conference on Space Debris,* ESA SP-672.

27. Klinkrad, H. (2006). *Space debris, models and risk analysis.* Berlin, Heidelberg, New York: Springer. ISBN: 3-540-25448-X.

28. Alfriend, K., Akella, M., et al. (1999). Probability of collision error analysis. *Space Debris, 1*(1), 21–35.

Chapter 7
The Satellite Control Facility

The Satellite Control Facility (SCF) is the element in the ground segment responsible to receive and manage all the telemetry (TM) received from the satellite and to transfer any telecommand (TC) to it. The SCF needs to initiate a contact through direct interaction with the TT&C station (baseband modem) using the project applicable space link protocol which is described in more detail in Sect. 7.2. As a TM dump will typically provide a vast amount of new TM values, the SCF needs to provide the means to perform a quick health check of all received TM and flag any *out-of-limit* parameter immediately to the operator. Furthermore, it needs to provide means to archive all received TM to allow a detailed trend analysis and error investigation.

The SCF is also responsible to maintain the satellite's onboard software and therefore needs to implement a module that is able to upload software patches or even entire new images to the satellite (see Sect. 7.3).

Once a satellite has moved from its LEOP phase into routine operations, routine contacts with a limited and repetitive set of TM and TC exchange will become the dominant regular ground-to-space interaction. To reduce the work load on operators, an automated satellite operations function is required which is described in Sect. 7.4.

Depending on the project specific needs for data encryption (e.g., encryption at TM or TC frame level), the necessary real time interaction with the security module in the ground segment needs to be performed and a possible architecture is described.

7.1 Architecture

Based on the main tasks described in the introduction, a simplified SCF architecture is presented in Fig. 7.1. As can be seen, the core part of the element is the TM/TC module which provides all the functionalities for the reception of telemetry and the

© The Author(s), under exclusive license to Springer Nature Switzerland AG 2023 119
B. Nejad, *Introduction to Satellite Ground Segment Systems Engineering*, Space
Technology Library 41, https://doi.org/10.1007/978-3-031-15900-8_7

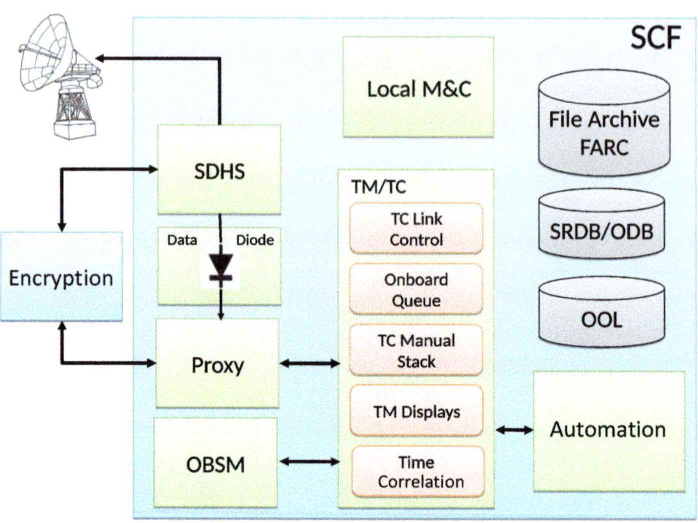

Fig. 7.1 Simplified architecture of the SCF. OBSM = Onboard Software Management, SDHS = Site Data Handling Set, OOL = Out-of-Limit, TM/TC = Telemetry and Telecommand Component

transmission of telecommands. As a modern TM database might comprise several hundreds of parameters (including derived ones), the TM display function is an important component to ensure an efficient presentation of this huge amount of information to the ground segment operator. The *time correlation* component has to translate the time stamps of all TM parameters from the satellite's onboard time system into the ground segment time system (e.g., UTC) and is described in more detail in Sect. 7.7.

The TC *manual stack* and TC *link control* components provide the operator the necessary tools to prepare a TC sequence to be uplinked to the satellite and to monitor its correct transmission and reception at satellite level.

The TM/TC component interfaces with the TT&C station in order to establish the ground-to-space link. This is realised through the *site data handling system* (SDHS) which incorporates the applicable ground-to-space communication protocol. For projects with specific requirements on encrypted satellite communication, all TM/TC traffic needs to be routed through a dedicated encryption unit. In order to avoid the encryption of SDHS server monitoring data, which is actually not related to the ground to satellite communication, a data-diode and proxy server architecture can be realised which is schematically shown in Fig. 7.1. The data diode is configured to only allow a traffic in one direction, i.e., *from* the SDHS *to* the proxy server and this path is only designated for data addressing the SDHS server monitoring like the status of hardware and software processes. No TM/TC data flow is allowed to take this route and must take the path through the encryption unit.

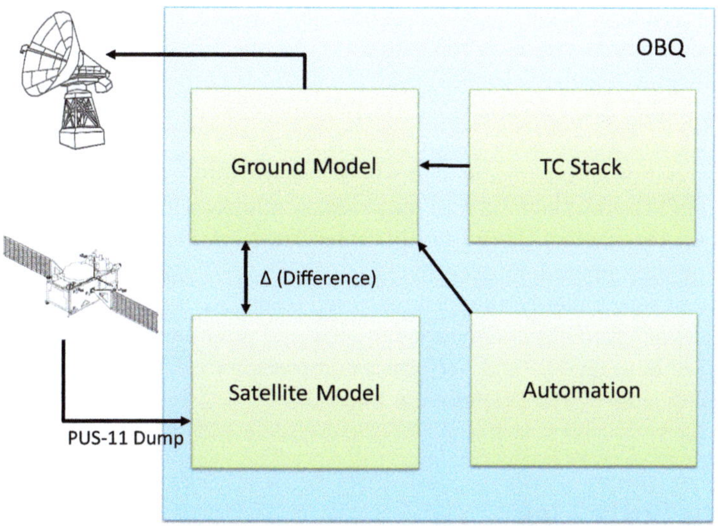

Fig. 7.2 SCF onboard queue model. PUS = Packet Utilisation Standard

The OBSM subsystem manages the upload of software patches or even full images to the satellite onboard computer and is described in more detail in Sect. 7.3.

The *automation* component provides the ability to execute operational procedures without the need of operator interaction and therefore requires full access to all functionalities of the TM/TC component. Automated satellite operations is a very important feature for satellite routine operations and even more relevant for projects that deploy satellite constellations (see Sect. 7.4).

The SCF needs to host the following set of databases which are shown in the upper right corner of Fig. 7.2:

- The *satellite reference database* (SRDB) and the *operational database* (ODB) contain the full set of TM and TC parameters defined in the onboard software, the definition of derived TM parameters, and the definition of TC sequences (see Sect. 9.2).
- The *out-of-limit* (OOL) database comprises for every TM parameter the expected range of values (e.g., nominal, elevated, and upper limits) which serves as a basis to determine whether an OOL event occurs that can be flagged to the operator accordingly (e.g., display of parameter in green, yellow, and red colour).
- The *file archive* (FARC) stores all received telemetry and makes it accessible for further analysis. Furthermore, the FARC stores the onboard software images and patches and makes them available to the OBSM component for further processing (see Sect. 7.3).

Many satellite projects make use of time-tagged TCs which are not intended to be executed immediately after reception but at a specified time in the future. This functionality has a high relevance for satellites in lower orbits with only short ground station visibility. The *onboard queue* (OBQ) model is used to keep track of time-tagged TCs and a high level architecture is shown in Fig. 7.2. The *ground model* keeps a record of all commands that where successfully released from the TM/TC component (TC stack) and the *spacecraft model* reflects what has been received and is stored on the satellite. Using the PUS Service 11 *command schedule reporting* function of the onboard software (refer to ECSS-E-70-41A [1]), the contents of the on-board queue can be dumped and used to update the spacecraft model. The OBQ model needs to provide a means to view and print the time-tagged commands contained in both models and perform a comparison to show potential differences which is indicated by the Δ symbol in Fig. 7.2.

7.2   The Space Link

The exchange of telemetry (TM)[1] and telecommands (TCs) is a fundamental concept that enables the remote operations of a satellite in space. Telemetry is the means for a satellite to transfer data to the ground segment and telecommands provides the means for the ground segment to command the satellite. The generation of TM requires the satellite onboard software to collect all relevant measurements from the various subsystems and to organise them into a format that is suitable for RF transmission to the ground segment. Similarly, the generation of TCs is a task of the ground segment (usually performed by the SCF) which need to be formatted for efficient and safe RF transmission to the satellite. The rules and methods to structure, format, and transmit TM and TC have to be clearly defined in a *space link standard*. In order to avoid the need to invent and specify project specific conventions, the use of international standards is a common and recommended practice since early space projects. This does not only reduce the development time and cost, but also opens the possibility to reuse existing processing hardware and software for both the satellite and the ground segment.

A widely applied space link standard is defined by the Consultative Committee for Space Data Systems or CCSDS, which is an organisation that was formed in 1982 by the major space agencies of the world and is currently composed of eleven member agencies, twenty-eight observer agencies, and over 140 industrial associates [2]. The CCSDS space link standard is outlined in various references that specify a suite of protocols, formats, rules, and services at different network layers

---

[1] The word *tele-metry* is derived from Greek roots: *tele* stands for remote and *metron* for measurement. In a broader sense, telemetry refers to any measurement which is performed at a remote location and has to be transferred via some means to an end user who takes care of any further processing.

OSI Model Layer	Space Communication Protocol	Major Features	CCSDS Standard Reference
Application	File Delivery Protocol (**CFDP**) Data & Image Compression Messaging (**AMS**)	• Asynchronous messaging service for mission modules • File delivery protocol • Lossless data, image, and spectral image compression	735.1-B-1 (ASM) 727.0-B-4 (CFDP) 121.0-B-2 (Data C.) 122.0-B-1 (Image) 123.0-B-1 (Spectral)
Transport	Space Communication Protocol Specification (**SCPS-TP**)	• Provision of end-to-end transport services; • Ability to use internet Transport Protocol (TP) and User Datagramm Protocol (UDP) over space data link.	714.0-B-2 (SCPS-TP)
Network	Space Packet Protocol (**SPP**) Encapsulation Service	• Space Packet routing via APID • Encapsulation Packets using IP	133.0-B-1 (SPP) 133.1-B-2
Data Link	Space Data Link Protocol (**SDLP**) for TM, TC, AOS, P-1; Communication Operation Procedure (COP); Addresssing Scheme	• Variable length TM/TC packets • Protocol Data Unit = Transfer Frame (TF); fix length for TM and AOS, variable length for TC and P-1; • Virtual Channels (VC): allow multiplexing of physical channel • Communications Operation Procedure or COP to retransmit lost/corrupted data	132.0-B-2 (TM/SDLP) 232.0-B-3 (TC/SDLP) 732.0-B-2 (AOS/SDLP) 211.0-B-5 (P-1/SDLP)
	Synchronisation & Channel Coding for TM, TC, AOS, P-1	• Synchronisatin & Channel coding schemes • Delimiting/synchronisation Transfer Frames	131.0-B-2 (TM) 231.0-B-2 (TC) 211.2-B-2 (P-1)
Physical	Radio Frequency & Modulation Systems		401.0-B-23 211.1-B-4 (P-1)

Fig. 7.3 CCSDS space communication protocol standards and mapping to the Open Systems Interconnection (OSI) basic reference Model. P-1 = Proximity-1, AOS = Advanced Orbiting System, SPP = Space Packet Protocol, CFDP = CCSDS File Delivery Protocol

which are summarised in Fig. 7.3, together with their main features. Furthermore, the relationship to the OSI *reference model*[2] is indicated.

One of the basic and central features of the CCSDS space link standard is the introduction of the *space packet protocol* (SPP) which defines the *space packet*[3] as a means for the user to transfer data from a source user application to a destination application via a *logical data path* (LDP) [4]. The structure of a space packet is shown in Fig. 7.4 and is applicable for both TC (referred to as TC source packet) and TM (referred to as a TM source packet). The following basic packet characteristics are worth noting:

- A space packet comprises a fixed size primary header and a variable data field.
- The transmission of packets can be at variable time intervals.
- The data field can contain an optional secondary header that allows the transmission of a time code and ancillary data in the same packet.

---

[2] The Open System Interconnection (OSI) basic reference model is a conceptual model for the characterisation of communication functions in IT and telecommunication systems. It was formally published by ISO in 1984 under ISO standard 7498 [3].

[3] The space packet is also referred to as the data unit of the Space Packet Protocol (SPP).

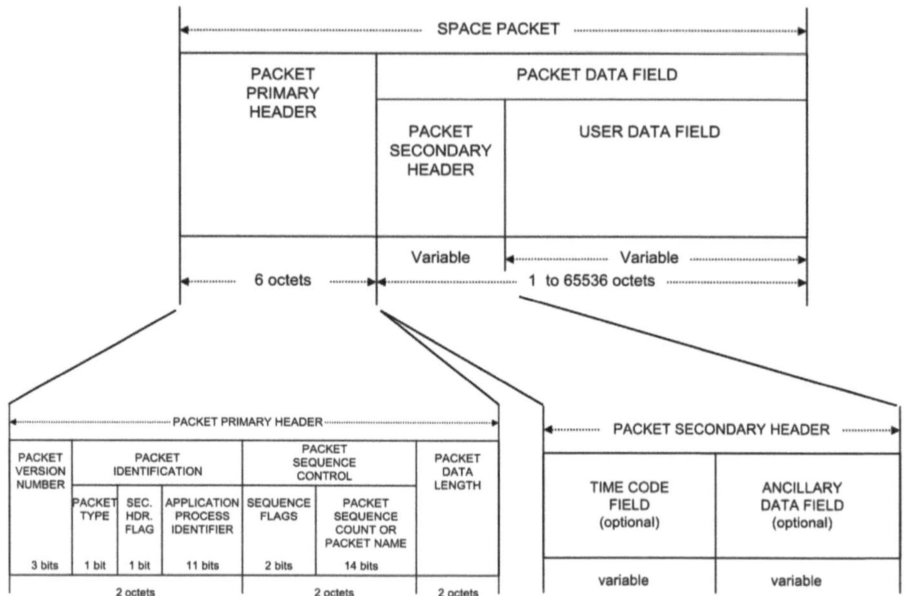

Fig. 7.4 Space Packet structure according to CCSDS-133-0-B-2 (refer to Fig. 4–1, 4–2, and 4–3 in [4]). Reprinted with permission of the Consultative Committee for Space Data Systems © CCSDS

- The packet header contains an *application process identifier* or APID which defines the routing destination through the logical data path (LDP).
- The packet sequence control fields provide a means to transmit sequencing information. This allows the correct processing of received packets even in case they are received in an incorrect sequence by the end user.

The next building block of the CCSDS space link is the *space data link protocol* (SDLP) which defines the *transfer frame* (TF) as the data unit for the transmission of source packets. The TC and TM transfer frames are schematically shown in Figs. 7.5 and 7.6 respectively and the detailed definition can be found in the corresponding standards (cf. CCSDS-232-0-B-3 [5] and CCSDS-132-0-B-2 [6]).

The TF contains a set of identifier fields which are referred to as the *transfer frame version number* (TFVN), the *spacecraft identifier* (SCID), and the *virtual channel identifier* (VCID). These identifiers are used in the context of another SDLP key feature which is referred to as *virtual channeling* (VC) concept that is used to multiplex several data streams onto one physical channel. This is shown in Fig. 7.7 where several source packets (with different APIDs) are multiplexed into one VC and various VCs are then again multiplexed into a master channel which is characterised by its own *master channel identifier* (MCID).

The next higher layer is referred to as the *synchronization and channel coding sublayer* and provides means for an optimised transfer of TFs over a physical space

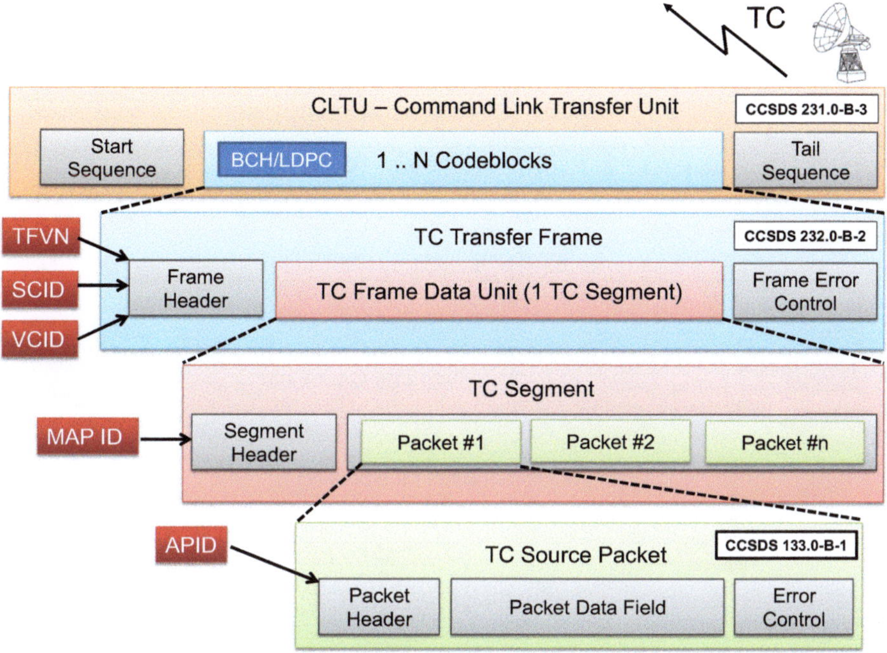

Fig. 7.5 The Command Link Transmission Unit (CLTU) definition. APID = Application Process ID, VCID = Virtual Channel ID, TFVN = Transfer Frame Version Number, SCID = Spacecraft ID, BCH = Bose-Chaudhuri-Hocquenghem, LDPC = Low Density Parity Check

channel. This layer defines error-control coding schemes, frame synchronisation methods (using attached sync markers), pseudo-randomization methods to improve the bit transition density of the datat stream,[4] and frame validation at the receiving end. The applicable standards CCSDS-131-0-B-3 [8] and CCSDS-231-0-B-3 [9] recommend specific channel coding schemes for TM (e.g., Reed-Solomon, convolutional, turbo, low-density parity check) and TC coding (e.g., BCH and LDPC). The final step is to combine the coded blocks into a *channel access data unit* (CADU) for the TM and a *command link transfer unit* (CLTU) for the TC. In contrast to the fixed size CADU of 1279 bytes, the CLTU has a variable size which depends on the number of code blocks which are encapsulated into it[5] (see Figs. 7.5 and 7.6).

---

[4] The frequency of bit transitions can influence the acquisition time required by the bit synchroniser which is part of the TT&C BBM and needed to demodulate the data stream from the subcarrier.

[5] One code block has size of 8 bytes. A maximum of 37 code blocks can be encapsulated into one CLTU.

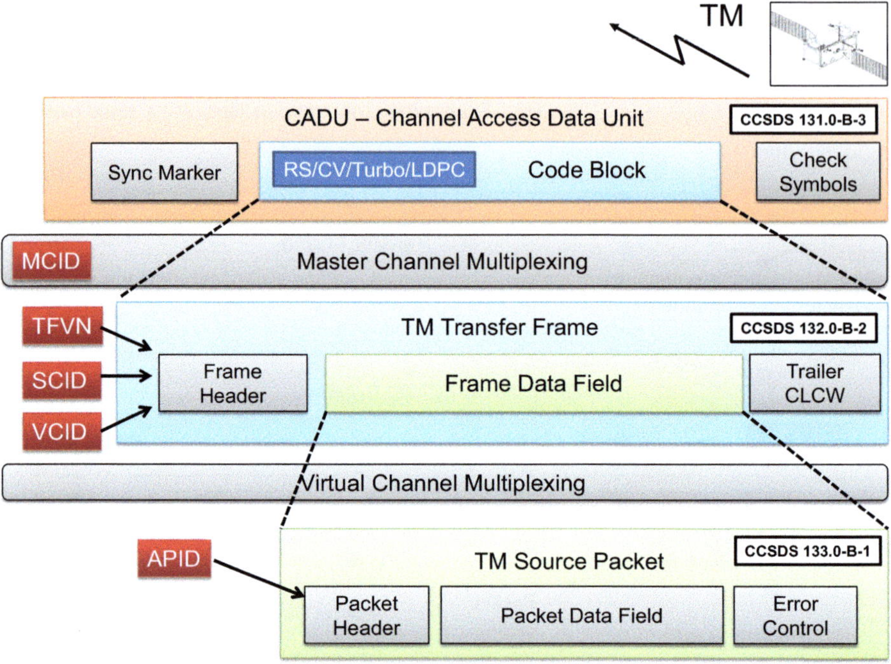

Fig. 7.6 The Channel Access Data Unit (CADU) definition. CLCW = Communication Link Control Word

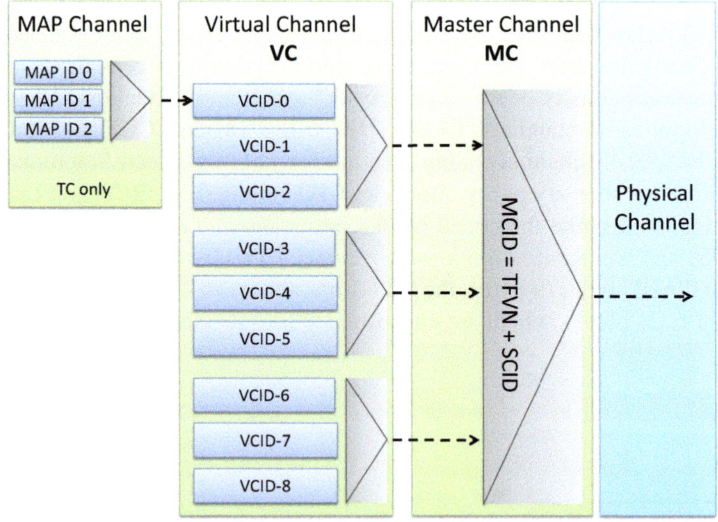

Fig. 7.7 CCSDS Space Data Link Protocol channel multiplexing concept. MCID = Master Channel ID, TFVN = Transfer Frame Version Number, SCID = Spacecraft ID [7]

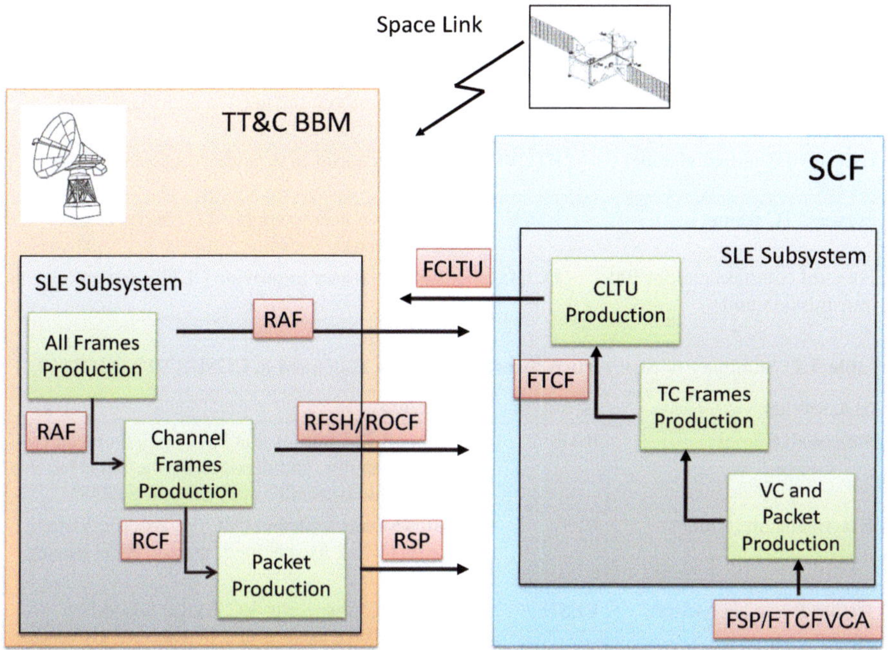

Fig. 7.8 CCSDS Space Link Extension (SLE) Services (refer to CCSDS-910-0-Y-2 [10] and Tables 7.1 and 7.2 for definition of acronyms)

7.2.1   SLE Services

Whereas format and structure of the TM and TC packets and frames were introduced in the previous chapter, this section explains the space link extension (SLE) services for the transportation of frames between the ground and space segment. An overview of the currently defined services is shown in Fig. 7.8 and a detailed description is provided in the SLE Executive Summary [10] and the references cited in there. The SLE services can be categorised according to their transfer direction into forward (from ground to satellite) and return (from satellite to ground) services with a short summary provided in Tables 7.1 and 7.2 respectively.

7.3   Onboard Software Management

The main task of the onboard software management system (OBSM) is to update the satellite software from ground. This encompasses both, the uplink of patches, defined as a partial replacement of software code, or even entire images that replace the existing software. The need for a software modification in first place can

Table 7.1 Summary of SLE **forward direction** services as defined in CCSDS-910-0-Y-2 [10]

SLE service	Acronym	Description
Forward space packet	FSP	Allows a single user to provide packets for uplink without the need to coordinate with different packet providers.
Forward TC virtual channel access	FTCVCA	Enables a user to provide complete VCs for uplink.
Forward TC frame	FTCF	Enables a user to supply TC frames to be transferred.
Forward communications link transmission unit	FCLTU	Enables a user to provide CLTUs for uplink.

Table 7.2 Summary of SLE **return direction** services as defined in CCSDS-910-0-Y-2 [10]

SLE service	Acronym	Description
Return all frames	RAF	Delivers all TM frames that have been received by the TT&C station (and decoded by the BBM) to the end user. This is usually the SCF in the GCS[a]
Return channel frames	RCF	Provides Master Channel (MC) or specific Virtual Channels (VCs), as specified by each RCF service user.
Return frame secondary header	RFSH	Provides MC or specific VC Frame Secondary Headers (FSHs), as specified by each RFSH service user.
Return operational control field	ROCF	Provides MC or VC Operational Control Fields (OCFs) channel, as specified by each ROCF service user.
Return space packet	RSP	Enables single users to receive packets with selected APIDs from one spacecraft VC.

stem from the discovery of a critical anomaly after launch or the need for new functionality. The main components of the OBSM module are shown in Fig. 7.9. The file archive or FARC has already been introduced before as an internal file repository that hosts the OBSW code as provided by the satellite manufacturer. The image model is an additional (optional) product which contains details on specific memory address areas and their attributes (e.g., whether an area can be patched or not). As OBSW images should be kept under strict configuration control, the Operations Preparation Facility or OPF described in Chap. 9 should be used as the formal entry point for the reception of OBSW updates and forward these to the FARC.

The three main OBSM components are shown in the central area of Fig. 7.9. The *image viewing* tool is used to load an image from the FARC and display its contents for detailed inspection prior to its uplink to the satellite. An example for an image representation in matrix form is shown in Fig. 7.10, where each line refers to

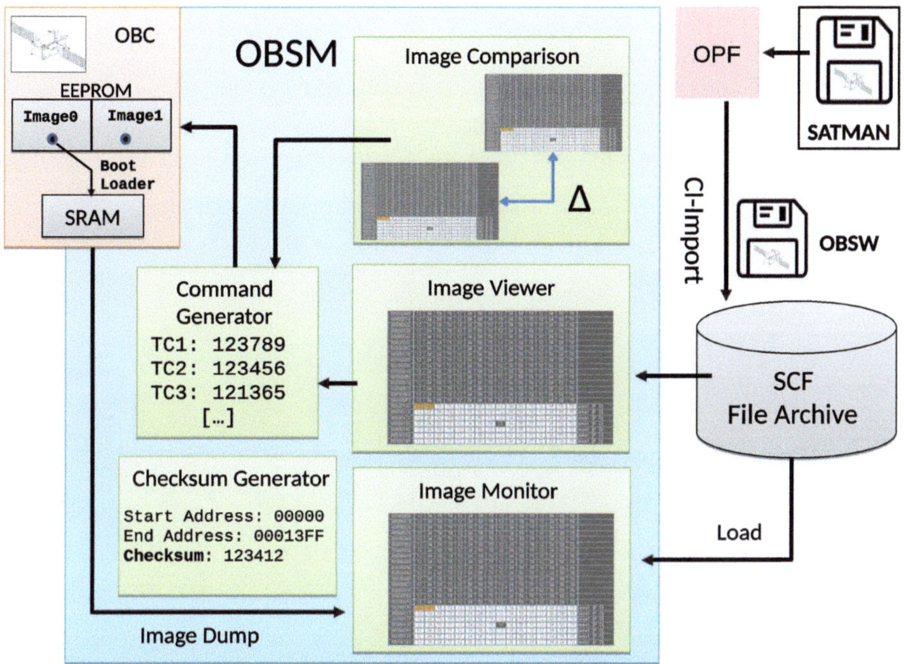

Fig. 7.9 The SCF onboard software management (OBSM) architecture

a dedicated memory address and each field in a column represents the contents of one information bit in that memory address. If a memory model is available, it can be used as an overlay to better display the memory structure.

The *image comparison* component provides a means to compare two images for differences, e.g. a previously uplinked one and a new one received.[6] If such differences are small, they can be used to generate a difference patch for transfer to the satellite which avoids the need to replace the entire image.

The *image monitoring* component has a similar function as the comparison tool but builds one of the images directly from live TM and does not load it from the archive.

After image viewing, comparison, or monitoring, the *command generator* converts the OBSW image into a sequence of TCs which are suitable for uplink to the satellite. After successful reception of the full set of TCs, the satellite onboard computer rebuilds the original image format. That image or patch is however not immediately used but stored in a dedicated area of the OBC memory which is sized to store two image versions at the same time (indicated by *Image-0* and *Image-1* in Fig. 7.9).

---

[6] In SCOS2K the terminology *uplink image* and *base image* are used to distinguish them.

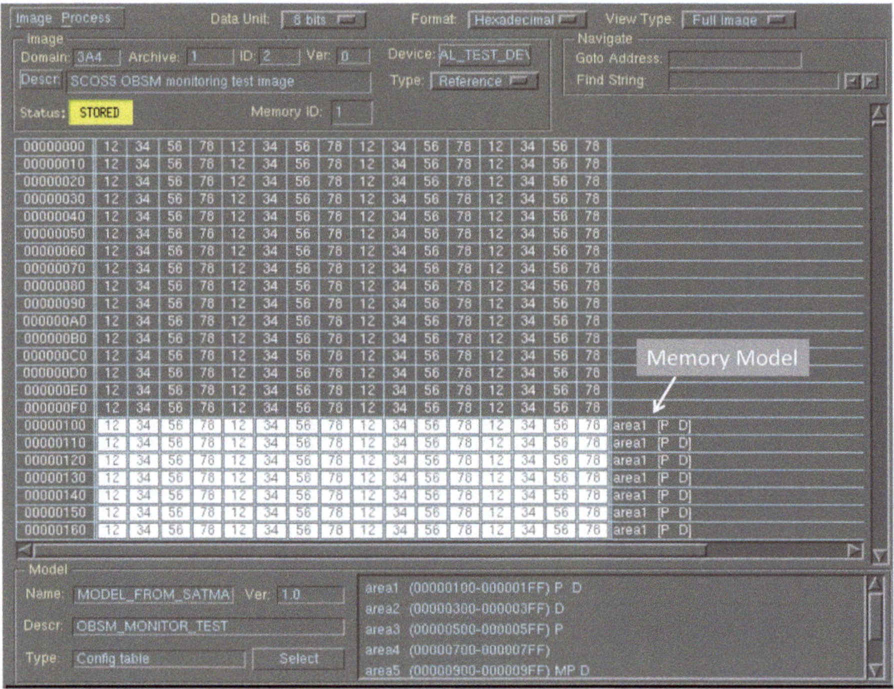

Fig. 7.10 The OBSM Image View in SCOS2K (© ESA/ESOC)

The non volatile memory referred to as *programmable read only memory* (PROM) or *electrically erasable* PROM (EEPROM) is used which retains its contents even if not powered on. This allows an in-depth verification of a new image prior to the replacement of the old one using a checksum comparison method to verify image correctness and completeness. A *checksum generator* uses a specialised algorithm to calculate checksums for both the uplinked and the on-ground images which can then be compared to each other. Only in case of equality of these checksums, the ground operators will instruct the OBC via a high priority commands (HPC)[7] to activate the new image which is then loaded into the static RAM (SRAM) by the boot loader.

---

[7] HPCs are a special set of TCs which can be executed without the need of the OBSW itself as they are routed directly in hardware to the dedicated unit that triggers emergency switching.

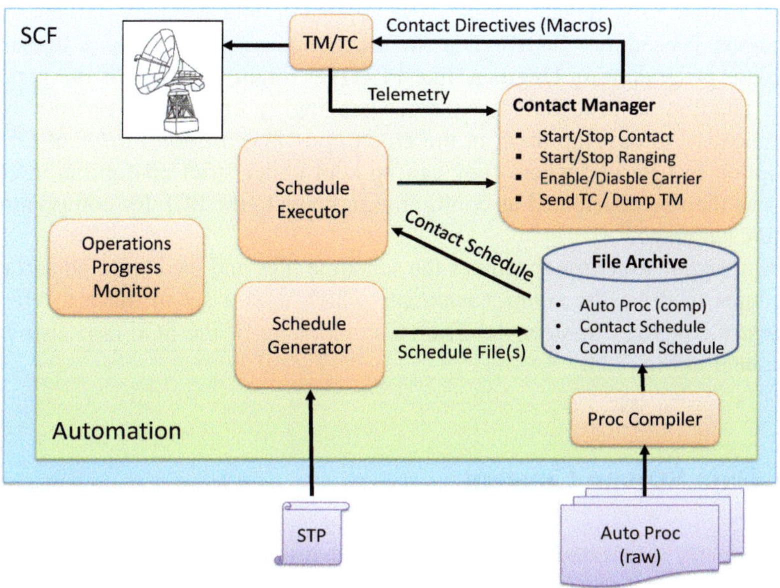

Fig. 7.11 Overview of the SCF Automation Module components. STP = Short Term Plan, Proc = Procedure

7.4 Automated Satellite Control

The automation component represents one of the most complex parts of the SCF and its detailed architecture can differ significantly from one project to the other. Only an overview of the most relevant components is shown in Fig. 7.11 which is inspired by the design of the automation component implemented in Galileo ground segment mission control system [11] and should therefore only be seen as an example. Independent of the detailed architecture, every automation component has to interface with the TM/TC module of the SCF in order to request TM dumps and uplink TCs to the satellites during a contact.

On the input side, the automation components receives the timeline of upcoming contacts and the required tasks and activities to be performed during each of the scheduled contacts. As this information is part of the mission plan, this input needs to be provided by the Mission Planning Facility (MPF) which follows the hierarchical planning philosophy as introduced in Chap. 8 with the Short-Term-Plan (STP) covering the most suitable time span for automated operations. The *schedule generator* is in charge to process the STP and generate the system internal schedule files which are stored in the local file archive (FARC) for internal processing.

Another important input for automation are the *automation procedures* which contain the detailed timeline (i.e., sequence of activities) of every contact with the exact reference to the applicable TM and TC mnemonics. These have to follow, similar to a manual contact, the applicable and formally validated Flight Operations Procedures (FOP). Due to the complexity of FOPs, a higher level procedure programming language like PLUTO[8] might be used for the generation of the automation procedures. As these are highly critical for operations, it is recommended to develop them in the *Operations Preparation Editor* of OPF (see description in Chap. 9) where they can be kept under strict configuration control. They can then be distributed as configuration items to the SCF for compilation into machine language.[9]

The *schedule executor* combines the schedule files and the compiled automation procedures to drive the *contact manager* which uses a set of macros referred to as *contact directives* to steer the TM/TC component of the SCF (see also TM/TC component in Fig. 7.1).

7.5   Data Stream Concept

The telemetry generated by the various subsystems of a satellite and can be categorised according to their generation time, storage onboard, and donwlink as follows:

- Realtime telemetry packets (HK-TM) are immediately downlinked when they are generated which requires a ground station contact to be established.
- Playback telemetry packets are generated at times when no ground station visibility or contact is available and must therefore be stored (or buffered) onboard the satellite so they can be transmitted at the next downlink opportunity. TM buffering is a widely used concept in space projects and requires adequate storage devices to be mounted inside the space assets. These are referred to as *mass memory and formatting units* (MMFU) or *solid state recorders* (SSR) and are highly complex devices which are optimised for large and fault tolerant storage capacity. They also have performant data input and ouput handling capabilities that allow read and write instructions at high data rates [13].
- Event driven TM packets are generated under specific circumstances (e.g., in case of an anomaly) and need to be stored and sent as playback TM to the ground segment as soon as possible, preferable at the next possible downlink opportunity.
- BBM buffer TM is stored inside the TT&C station after reception on ground and used in case of a temporary loss of network connection which prohibits any transfer to the main ground segment. Once the connection is reestablished, the buffer TM can be sent to the SCF to avoid any data loss.

---

[8] PLUTO stands for *Procedure Language for Users in Test and Operations* and is an operational programming language standardised in ECSS-E-ST-70-32C [12].

[9] The need for a dedicated procedure compiler depends on the system architecture but is usually required if the automation design is based on a higher level procedure languages like PLUTO.

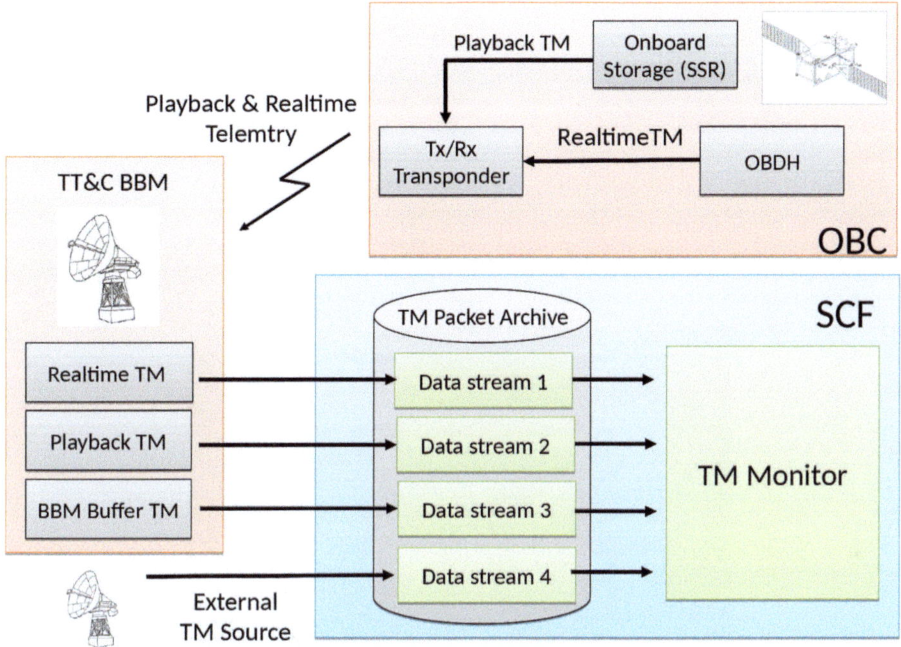

Fig. 7.12 The Datastream concept in SCF. SSR = Solid State Recorder, OBDH = Onboard Data Handling System, BBM = Baseband Modem, OBC = onboard computer.

Both real time and buffered TM packets need to be downlinked from the satellite to the ground segment via the same physical RF channel and are therefore multiplexed into one data stream. In order to separate the different type of TM flow at the receiving end again, the *virtual channel* concept is provided by the space data link protocol (SDLP) described earlier (see also Fig. 7.7). This allows the SCF *data stream* functionality to segregate the multiplexed TM into the different streams again and store them in dedicated areas of the TM *packet archive*. The TM monitoring can then access and display the various streams independently which is schematically shown in Fig. 7.12.

A specific data stream can also be reserved for TM received by project external infrastructure which is only booked for a certain project phase. A typical example would be the use of additional TT&C stations from external station networks during a LEOP phase to satisfy more stringent requirements for ground station coverage.

7.6 Telemetry Displays

Telemetry displays are the most frequently used applications of the TM/TC component as they provide the means for an operator to inspect and analyse the received

AND#01023_OCM

Sample Time	Reference	Description	Value	Unit
2009.105.13.07.23.023	THERM10033	Tank Pressure	22.25	bar
2009.105.13.07.23.023	AOCS0033	Satellite Roll	1.2563	deg
2009.105.13.07.23.023	AOCS0034	Satellite Pitch	51.3463	deg
2009.105.13.07.23.023	AOCS0034	Satellite Yaw	98.5632	deg
2009.105.13.07.23.023	AOCS0038	Thruster 1 Flow On	10.23	Sec
2009.105.13.07.23.023	AOCS0039	Thruster 2 Flow On	0.23	Sec
2009.105.13.07.23.023	AOCS0040	Thruster 3 Flow On	11.54	Sec
2009.105.13.07.23.023	AOCS0041	Thruster 4 Flow On	0.23	Sec

GD#01453_PRESS

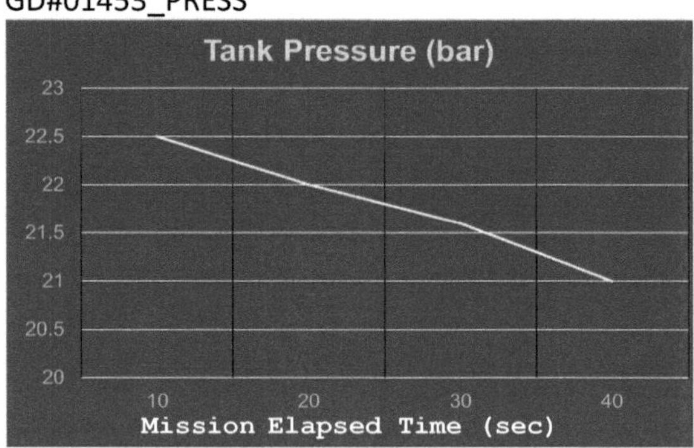

Fig. 7.13 Examples of an alphanumeric display (AND) and a graphical display (GD) in SCF.

telemetry parameters. In live mode, the operator inspects new TM parameters as soon as received and processed on ground. Alternatively, an operator might use TM displays to scrutinise historical satellite data by accessing archived TM files. The majority of TM displays deployed in the SCF can be affiliated to one of the following categories:

- Alphanumeric displays or ANDs provide a representation in a simple tabular format. Each line of the table displays one TM parameter next to a time stamp (usually UTC). Furthermore, the reference or name of the parameter, a short description (e.g., tank pressure), and the TM value with its correct physical dimension are displayed. An example of an AND is depicted in the upper panel of Fig. 7.13. The SCF should provide the possibility to tailor the content of an AND according to the operational needs. Dedicated ANDs with parameters specific to a satellite subsystems are usually preconfigured and can be made available to the specialised subsystem engineer. It also recommended to define ANDs at satellite system level which comprise parameters of various subsystems which are relevant during a specific operational phase. An example is a system AND prepared for the monitoring of the proper execution of an orbit correction manoeuvre which should combine TM parameters from the AOCS subsystem (e.g., attitude and thruster activity), power subsystem (e.g., battery capacity), and the satellite transponder (e.g., transmitter status).

- A different type of display is a graphical display which shows the time history of a selected TM parameter as a two dimensional graph (see lower panel of Fig. 7.13). This type of representation is especially useful if the time history and trend of a TM parameter needs to be monitored.
- A third type of display is referred to as a mimic display and uses geometric shapes (usually squares, triangles, and interconnection lines) that represent relevant satellite components. A colour scheme can be used to show the operational state of a component. An example could be a change from red to green, if a transmitter is activated or a celestial body enters into the FOV of a sensor. The operational state can be derived from a single or even a set of TM values and the corresponding background logic needs to be preconfigured in order to make a mimic meaningful. The look and feel of mimics displays are comparable to the MMI of the Monitoring and Control Facility which are described in Chap. 10 and shown in Fig. 10.3.

7.7 Time Correlation

Every satellite possess an onboard clock which is able to maintain a spacecraft reference time that is usually referred to as *spacecraft elapsed time* or *onboard time* (OBT). This time is based on a free running counter that starts at an arbitrary initial epoch and is used to time tag all telemetry events on the satellite. The onboard clock itself is usually driven by a numerically controlled oscillator (NCO) which either runs on its own or is steered (and synchronised) to an external time pulse, generated by a more precise time source (see upper satellite box of Fig. 7.14). Such an external satellite time source could be a navigation payload receiving a GNSS time signal or a highly precise atomic clock which us usually deployed on every navigation satellite. The onboard computer has no knowledge of absolute time, which means that all TM transmitted to ground must be time stamped in OBT. A conversion from OBT to an absolute time scale must therefore be performed in the ground segment.[10] Once a TM frame is received on ground and carrier demodulated by the TT&C (BBM), it is stamped with an absolute time label referred to as *Earth received time* (ERT). Once transmitted to the SCF, a proper time correlation between absolute time (e.g., UTC) and OBT is performed by correcting the ERT time stamp for the elapsed propagation time δ_i and the transmission delay time τ_i at time t_i. The value τ_i reflects the time (or latency) to process a TM frame and packet on ground. The relation to correct ERT therefore reads

---

[10] Note that satellites in low Earth orbit carrying a GNSS receiver payload could also perform absolute time stamping onboard.

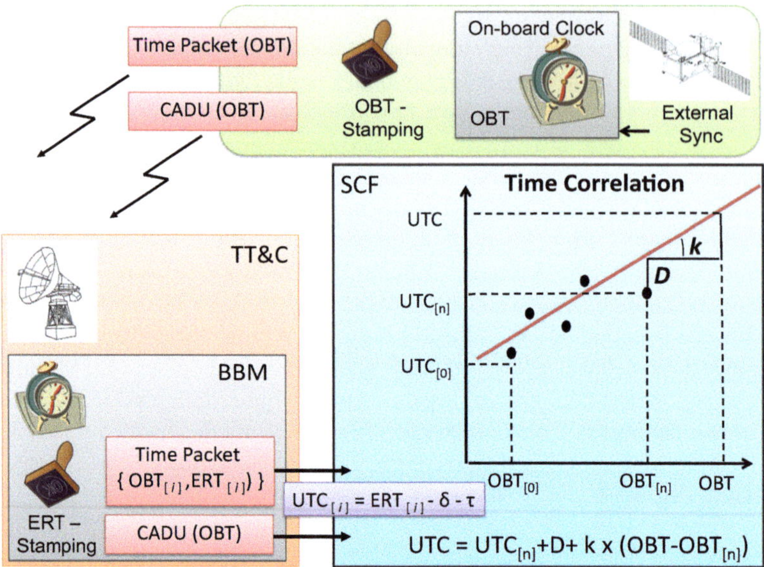

Fig. 7.14 The concept of time correlation performed by the SCF. OBT = On-board Time, ERT = Earth Received Time, UTC = Universal Time Coordinated.

$$UTC_{[i]} = ERT_{[i]} - \delta_i - \tau_i \qquad (7.1)$$

which allows to *correlate* a time couple $\{OBT_{[i]}, UTC_{[i]}\}$ at each epoch t_i. The absolute time system UTC is used as an example here being a widely used system in many ground segments. It could however also be replaced by a GNSS based system like GPS time or Galileo System Time (GST) which are readily available today. As the assumed δ and τ in Eq. (7.1) might differ from the real values by a different offset at each time t_i, the time correlation function in the SCF applies a linear regression module on a specified set of collected time packets that consists of time couples $\{OBT_{[1,...,n]}, UTC_{[1,...,n]}\}$ as indicated by the linear regression line in red in Fig. 7.14. The result yields the correlation coefficients D and k which can be used to convert any TM parameter time stamp from OBT to UTC using the following simple equation

$$UTC = UTC_{[n]} + D + k \times (OBT - OBT_{[n]}) \qquad (7.2)$$

where $UTC_{[n]}$ and $OBT_{[n]}$ can be any arbitrarily chosen initial time couple. As the linear regression is performed on a limited set of time couples only, the resulting time correlation parameters will have limited accuracy and validity. The time correlation accuracy is defined as the distance of a measurement point to the linear regression line and needs to be monitored to not exceed a defined limit, otherwise D and k need to be recomputed based on a new set of time couples.

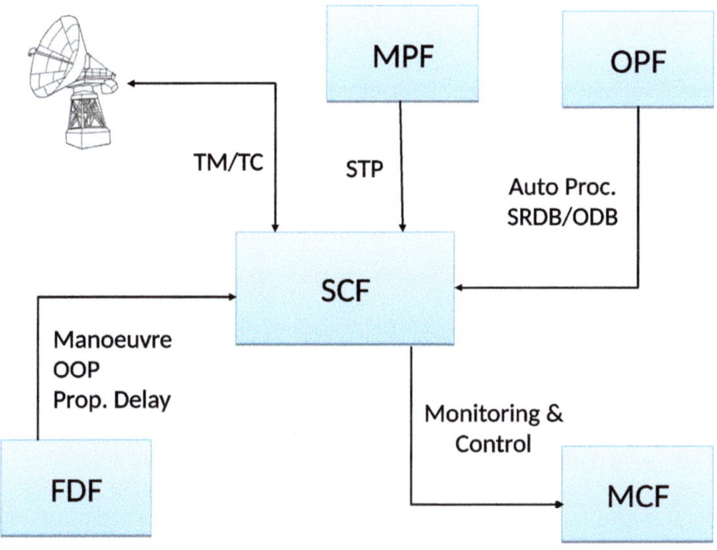

Fig. 7.15 Overview of the most important SCF interfaces to other GCS elements.

7.8 Interfaces

The most relevant interfaces between the SCF and the other ground segment elements are shown in Fig. 7.15 and described in the bullets below:

- **FDF** (see Chap. 6): the SCF needs to receive from the FDF parameters for flight dynamics related TCs. An example is a TC to execute an orbit control manoeuvres which usually requires the thrust duration, propulsion system specific parameters, and manoeuvre target attitude angles. Furthermore, the FDF needs to provide the command parameters to update the satellite OOP and send the propagation delay file that specifies the elapsed propagation time δ for each satellite to ground station link geometry (refer to section describing the time correlation).
- **MPF** (see Chap. 8): the automation component of the SCF requires as input the Short Term Plan (STP) which needs to be provided by the Mission Planning Facility. The STP must contain a detailed contact schedule and command sequences for the upcoming days that can be used for the automated contacts.
- **OPF** (see Chap. 9): the Operations Preparation Facility (OPF) provides the SCF two important inputs: the satellite reference database (SRDB) which contains the full set of satellite specific command and telemtry parameters, and the automation procedures described earlier.
- **MCF** (see Chap. 10): the interface to the Monitoring and Control Facility ensures the central monitoring of the SCF element hardware and processes and provides a basic remote commanding functionality to change element operational modes.

References

1. European Cooperation for Space Standardization. (2003). *Ground systems and operations - Telemtry and telecommand packet utilization.* ECSS-E-70-41A
2. The Consultative Committee for Space Data Systems. (2022) . https://public.ccsds.org/about/. Accessed 11 March 2022.
3. International Organization for Standardization. (1994). *Information Technology — Open Systems Interconnection — Basic Reference Model — Conventions for the Definition of OSI Services.* ISO/IEC 10731:1994.
4. The Consultative Committee for Space Data Systems. (2020). *Space Packet Protocol. CCSDS 133.0-B-2.*
5. The Consultative Committee for Space Data Systems. (2015). *TC Space Data Link Protocol. CCSDS 232.0-B-3.*
6. The Consultative Committee for Space Data Systems. (2015). *TM Space Data Link Protocol. CCSDS 132.0-B-2.*
7. The Consultative Committee for Space Data Systems. (2014). *Overview of Space Communication Protocols.* Informational Report (Green Book), CCSDS 130.0-G-3.
8. The Consultative Committee for Space Data Systems. (2017). *TM Synchronization and Channel Coding. CCSDS 131.0-B-3.*
9. The Consultative Committee for Space Data Systems. (2017). *TC Synchronization and Channel Coding. CCSDS 231.0-B-3.*
10. The Consultative Committee for Space Data Systems. (2016). *Space Link Extension Services - Executive Summary. CCSDS 910.0-Y-2.*
11. Vitrociset. (2010). *Galileo SCCF Phase C/D/E1 and IOV, SCCF Automation and Planning Component, Software Operations Manual (SW5). GAL-MA-VTS-AUTO-R/0001.*
12. European Cooperation for Space Standardization. (2008). *Test and operations procedure language. ECSS-E-ST-70-32C.*
13. Eickhoff, J. (2012). *Onboard computers, onboard software and satellite operations.* Heidelberg: Springer.

Chapter 8
The Mission Planning Facility

The mission planning discipline has already been briefly introduced in Sect. 4.1.7 and aims to efficiently manage the available ground segment resources, consider all known constraints, and resolve all programmatic conflicts in order to maximise the availability of the service provided to the end user. The Mission Planning Facility (MPF) represents the ground segment component that implements this functionality and depends on planning inputs shown in Fig. 4.2. A more detailed description of the functional requirements is given below.

- The reception, validation, and correct interpretation of *planning requests* (PRs) is the starting point of any planning process. Such PRs can originate from an operator sitting in front of a console or stem from an external source which could be either an element in the same ground segment or an external entity. External PRs are of special relevance in case a *distributed planning* concept is realised which is outlined in more detail in a later part of this chapter. Once a PR has been received, it should always be immediately validated to ensure correctness in terms of format, meaningful contents, and completeness. This avoids problems at a later stage in the process when the PRs contents is read by the planning algorithm.
- The reception and processing of the flight dynamics event file containing the ground station visibilities is an essential task that allows the correct generation of the contact schedule. Furthermore, the event file provides epochs which are needed for the correct placement of tasks which require specific orbit event related conditions to be fulfilled. Examples could be the crossing of the ascending nodes, the orbit perigee/apogee, satellite to Sun or Moon co-linearities, or Earth eclipse times.
- The planning algorithm (or scheduler) must be able to consider all defined constraints and, in case of conflicts, be able to solve them autonomously. Such an automated conflict resolution can only be performed through the application of a limited number of conflict solving rules which have been given to the algorithm. If these rules do not allow to find a solution, the software must inform the

B. Nejad, *Introduction to Satellite Ground Segment Systems Engineering*, Space Technology Library 41, https://doi.org/10.1007/978-3-031-15900-8_8

operator and provide all the necessary information as output to facilitate a manual resolution.

- Once the plan has been generated, it needs to be distributed to all end users in the expected format. Mission planning products can comprise a contact schedule, a mission timeline, or an executable plan.
- The processing of feedback from the mission control system (SCF) about the status of the execution of planned tasks and activities allows to update the planning time line. This update is an important input that should be considered in the subsequent planning cycle. As an example, a planned task that could not be executed would need to be replanned to a future slot which would impact the future planning schedule.

8.1   Architectural Overview

Based on the outlined functionalities, a simplified mission planning architecture is shown in Fig. 8.1 where the direction of the planning flow is indicated via an arrow pointing from the lower left to the upper right corner. The *planning request manager* is ready to receive, validate, and process any incoming PR at any time and should therefore be implemented as a background process running continuously (e.g., as a daemon process). The *contact scheduler* implements the scheduling algorithm which must be able plan satellite contacts that accommodate all planning requests within the available ground station visibility times. Whereas it might be evident that a contact must always be placed within the start and end times of a ground

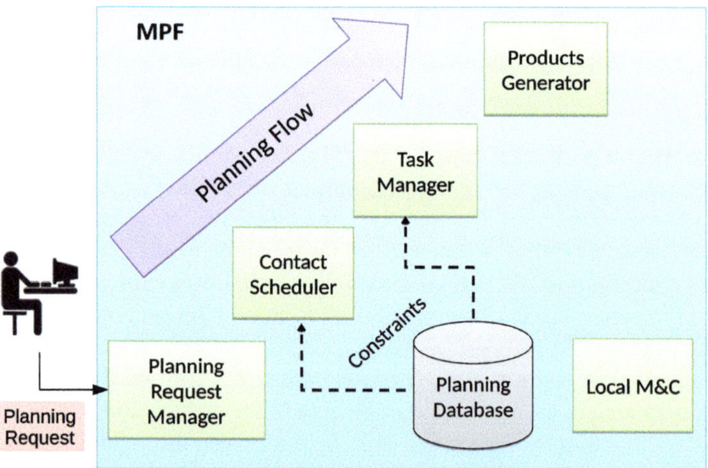

Fig. 8.1 Simplified architecture of the Mission Planning Facility (MPF) indicating the planning flow direction

station visibility, it must be kept in mind that additional contact rules and constrains can shorten the actually planned contact duration. The output of the scheduling algorithm is a valid contact plan which can be provided to the *task manager* in order to place the relevant tasks and activities in their correct sequence. In the final step, the *products generator* creates all the planning products in the required format and file structure and distributes them to the end user.

8.2 Planning Concepts

The aim of a planning concept is to describe the overall planning sequence and philosophy. This should already be defined and documented in the early project phases as it can impact the architecture and interfaces of the MPF. Only the most applied concepts are described here using the standardised terminology from CCSDS-529.0-G-1 [1]. All the described planning concepts should not be considered as exclusive and can in principle be combined.

Hierarchical planning is a concept that aims to structure the planning task into several cycles with each covering a different time span (planning window) with a different level of detail. Whereas the higher-level cycles deal with longer time spans, the lower-level ones go into more detail and granularity (see Fig. 8.2). The most common planning windows are the long-term (months up to years), medium-term (weeks up to months), and short-term (days up to weeks) cycle with the corresponding long-term-plan (LTP), mid-term-plan (MTP), and short-term-plan (STP) attached to them. A typical example for a task considered as part of the LTP is a launch event which is usually known months ahead of time. A task considered in the MTP could be the execution of an orbit correction manoeuvre which should be known weeks ahead of time. The update of the satellite onboard orbit propagator might be a weekly task and can therefore be considered as part of the STP.

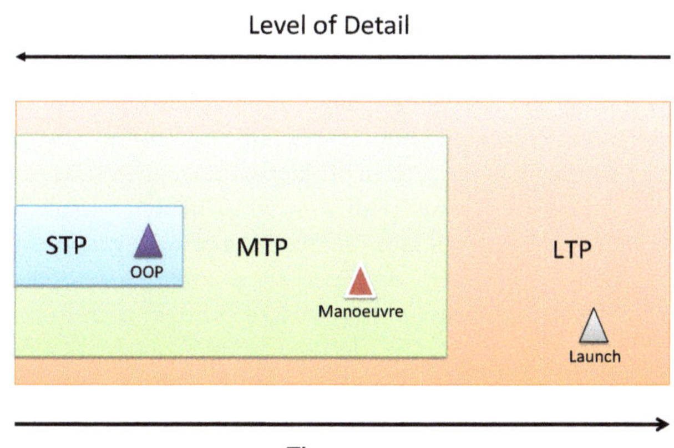

Fig. 8.2 Hierarchical planning concept with three different planning cycles: STP = Short-Term-Plan, MTP = Medium Term Plan, LTP = Long Term Plan. The time span increases from left to right, whereas the level of detail increases in the exact opposite direction

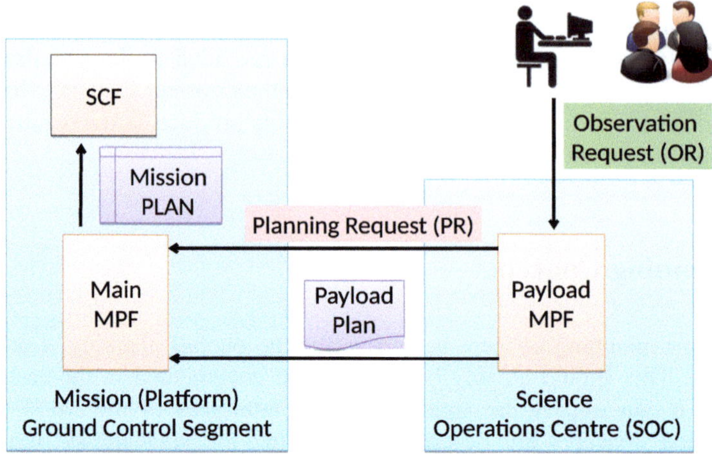

Fig. 8.3 Concept of *distributed planning* showing the planning functionality distributed among two different entities, one specialised for the payload and the other one for the overall satellite and mission planning. SCF refers to the Satellite Control Facility as described in Chap. 7

In addition to the duration and level of detail, a hierarchical planning concept needs to define the interaction rules for the areas in which adjacent cycles overlap. A meaningful rule is to give tasks defined in longer planning cycles a higher priority compared to the ones from shorter cycles. The reasoning behind this is that planners of longer cycles consider project needs and constraints that must not necessarily be known to the planners of shorter cycles. Applying this to the previously mentioned example, one could assume that a launch event is planned in a different forum and by a different group of people compared to a routine satellite event like the update of the OOP. Following this idea, the generation of the STP has to respect all the planned tasks and activities of the MTP and LTP and place any new tasks around these.

The terminology of *distributed planning* is used in case the planning process is not concentrated in a central facility but distributed among several different planning entities. Such a separation could be driven by the availability of specific resources like skill sets or specialised facilities, or result from the distribution of responsibilities for certain project assets to different centres, groups or even agencies (e.g., payload versus platform, rover/lander versus orbiter, etc.). The concept of distributed platform and payload planning is graphically depicted in Fig. 8.3 and is frequently applied in satellite projects that carry complex scientific payloads (e.g., space telescopes, rover missions, or orbiters with multiple sensor platforms). In this case, the PRs are usually derived from *observation requests* (OR) that were initiated by the scientific community. A dedicated *science operations centre* (SOC) could be used to establish the interface to the external community in order to collect, analyse, and coordinate all incoming ORs. The coordination of different observations might not always be a trivial task due to the potential request of contradictory satellite attitudes by different ORs aiming to achieve an optimum observation condition. Once the payload planning has been finalised, a set of PRs or a even a consolidated payload plan that contains all valid and de-conflicted ORs can be sent to the main planning entity which coordinates this input with PRs related to the satellite and ground segment maintenance. The need to merge inputs from various planning entities into one consolidated plan is the main characteristic and also challenge of the distributed planning concept. In case of conflicting requests, a predefined priority list should be applied. An example rule could be that PRs related to satellite maintenance (e.g., OOP update or orbit change manoeuvre) have a higher priority than ground segment maintenance activities (e.g., software upgrade). The latter one again has a higher priority than a scientific observation request, which can be scheduled at a later time.

In an *iterative planning* concept the planning cycle is started whenever new input arrives. This could be a new PR (e.g., new observation request), a new constraint (e.g., sudden downtime of a ground station), or a new orbit event (e.g., sensor blinding by Sun or Moon). Due to the dynamic nature of satellite missions, it is highly recommended to always incorporate an iterative planning philosophy up to a certain degree. It must be clearly defined at what point a new planning cycle needs to be initiated, and performance requirements should specify the expected convergence time to obtain an updated mission plan.

Finally *onboard planning* and *goal based planning* are briefly mentioned. The onboard planning concept is based on the satellite onboard scheduling abilities that use conditions or events which the satellite can detect autonomously without the need for ground segment interaction (e.g., eclipse, sensor blinding, sensor FOV events). Goal based planning refers to a concept based on PRs serving as the primary input for the planning function to schedule the correct tasks and activities in order to process the PR.

8.3   The Mission Planning Process

The *mission planning process* (MPP) defines the steps required to generate a conflict free mission plan that contains the exact times for all tasks to be executed by the satellite platform, its payload, or the ground segment. Figure 8.4 shows the basic steps and sequence of a typical MPP starting with the processing of all planning inputs like the orbit events and the full list of planning requests (PRs). In the next step the contact schedule (CS) is generated which takes into account the available ground station visibilities from the event file and the project specific contact rules (or constraints) indicated by the *contact rules* box. It is important to distinguish between the duration of a visibility which is given by the time span during which a satellite has an elevation of at least 5 deg (or higher) above the ground station's

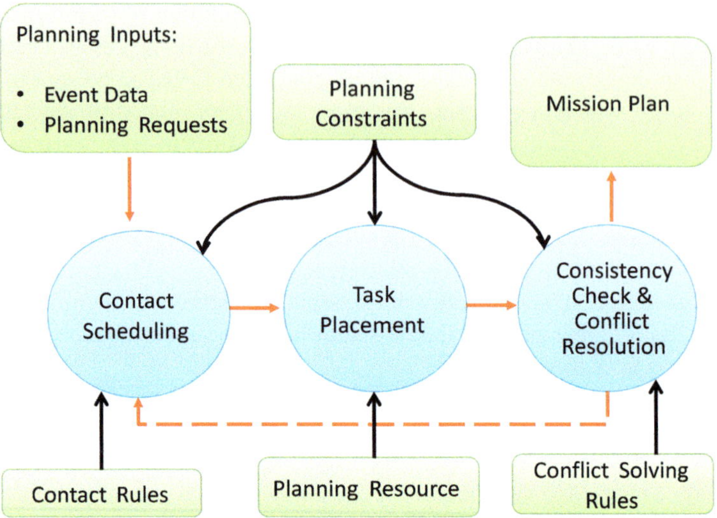

Fig. 8.4 Overview of Mission Planning Process (MPP). Inputs and outputs to the process are depicted as boxes whereas processes itself as circles

horizon,[1] and the contact duration itself which can be much shorter, as it is driven by the set of tasks that need to be executed. The start and end times of a contact are termed *acquisition of signal* (AOS) and *loss of signal* (LOS) and refer to the actual ground station RF contact to the spacecraft transponder for the exchange of TM/TC and ranging (see Fig. 8.5). Typical contact rules comprise a minimum and maximum contact duration, constraints on gaps between two consecutive contacts (to the same satellite in a constellation), or requirements on the minimum number of contact per orbit revolution (contact frequency).

The next step in the planning process is the placement of tasks and their related activities within the available contact slots which is shown in the lower portion of Fig. 8.5. The task placement can be either done based on absolute times specified in the planning request like the earliest or preferred start time $t_{earliest}$ or t_{pref} shown for "Task 4", or via a relative time (or offsets) specification Δt defined with respect to a specific orbital event like the crossing of the perigee (see "Task 1") or the start of a contact (see "Task 2"). The relative task placement can be automated by the planning algorithm using the definition of a *relative-time task trigger* that is able to detect the occurrence of a specific event.

Instead of a specific orbit event, the successful execution of a task itself could also be a condition for the start of a subsequent one. This can be done via the definition of so called *task-to-task triggers* as shown for "Task 3". The placement of tasks might

---

[1] Satellite ground station visibilities can span from a few minutes for orbits at low altitude up to several hours for high altitude orbits (e.g., MEO) and ground stations close to the equator.

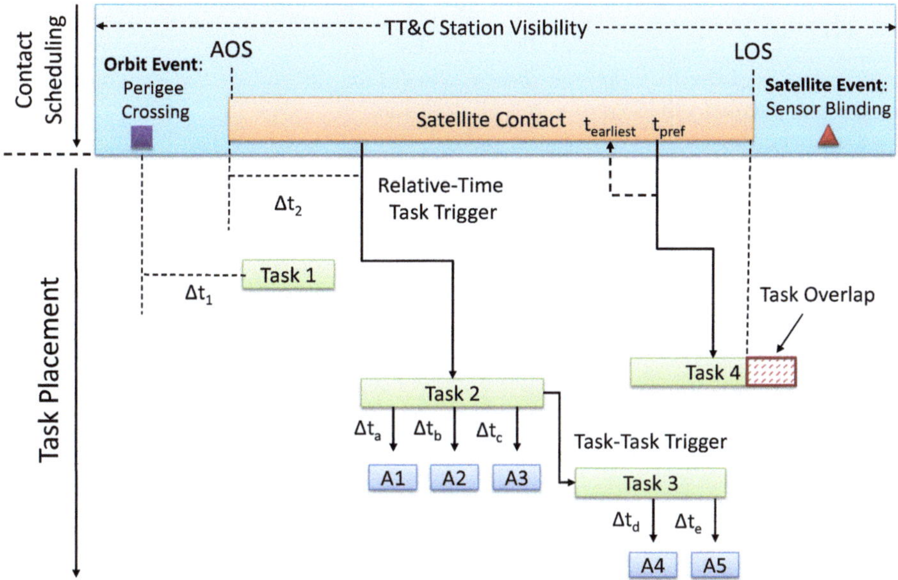

Fig. 8.5 Contact scheduling and Task Placement outcomes (refer to text for detailed explanation)

also be depend on the availability of specific planning resources which could be the availability of a certain battery load level[2] or a minimum amount of propellant required for the execution of an orbit change manoeuvre. The dependency of a task on a specific resource is an important constraint for its placement and therefore requires as input an up-to-date state of the involved resources that must be acquired from a dump of recent housekeeping telemetry.

The last step in the MPP is the consistency check that ensures that none of the placed tasks violates any planning constraint. An example for such violation is shown in Fig. 8.5 for the placement of "Task 4", where the duration to execute this task in combination with the intended preferred start time t_{pref} would imply its end time to be located after the contact LOS time (indicated by the "Task Overlap" label). In case such a violation is detected, a possible resolution must be attempted which in the given example could be achieved by advancing the actual start time of the task to the earliest allowed start time $t_{earliest}$. This example clearly demonstrates the benefit to always define both an earliest and a preferred start time in every PR as this provides additional flexibility to resolve scheduling conflicts.

The definition of conflict solving rules is an important input to the planning process as it allows the scheduling algorithm to attempt a first conflict resolution without the immediate need for human interaction. In case no solution can be found

---

[2] The battery power is usually much lower after the satellite exits an eclipse period which might preclude the execution of power intense satellite tasks.

by simply shifting the start times, a change in the overall scheduling approach might be needed. For such cases, the conflict solving rules should also provide clear guidelines on how to relax scheduling constraints and allow the generation of a consistent contact schedule in a second iteration.[3]

Once the consistency check has passed, the mission plan can be finalised and released. One important aspect of constellation type missions is the need to perform mission planning for all satellites simultaneously, which also implies that the mission plan must contain a list of contacts and tasks for all satellites. Some constraints or resource limitations (e.g., the down time of a TT&C station due to a maintenance activity) might have a potential impact on the overall schedule and should therefore be categorised as system wide resources.

8.4 Contact Scheduling

The complexity of the contact scheduling process is strongly driven by the actual mission profile. As an extreme example, for a project that only operates one single satellite, one single ground station, and implements an orbit with a repeating ground track,[4] the contact schedule will be very simple and highly repetitive. In contrast to this, a constellation type mission with several satellites (e.g., a navigation type constellations can have 30 or more spacecraft in service), a large TT&C ground station network deployed around the globe, and a complex set of contact rules (e.g., minimum of one contact per orbit revolution, need to schedule back-up contacts, need on specific contact duration, etc.), the generation of the contact schedule requires a highly complex algorithm.

The difference between the ground station visibility being the physical satellite location of 5 deg (and higher) above the ground station horizon and the actual contact time defined by AOS and LOS has already been briefly introduced in the previous section. The following contact scheduling parameters are highly relevant for the computation of contact times and must therefore be clearly defined.

- The minimum contact duration between AOS and LOS is driven by the amount of time a TT&C station requires to establish an RF contact. This parameter should be defined based on the TT&C station performance[5] to avoid that the contact scheduling process creates contacts which cannot be realised from an operation point of view.

---

[3] This iterative approach is indicated by the dashed arrow pointing from the consistency check circle to the contact scheduling circle in Fig. 8.4.

[4] Orbit geometries with ground tracks that follow a fixed repeat cycles are commonly chosen for Earth observation type missions.

[5] E.g., the minimum time to activate all the components on the TT&C transmission and reception paths.

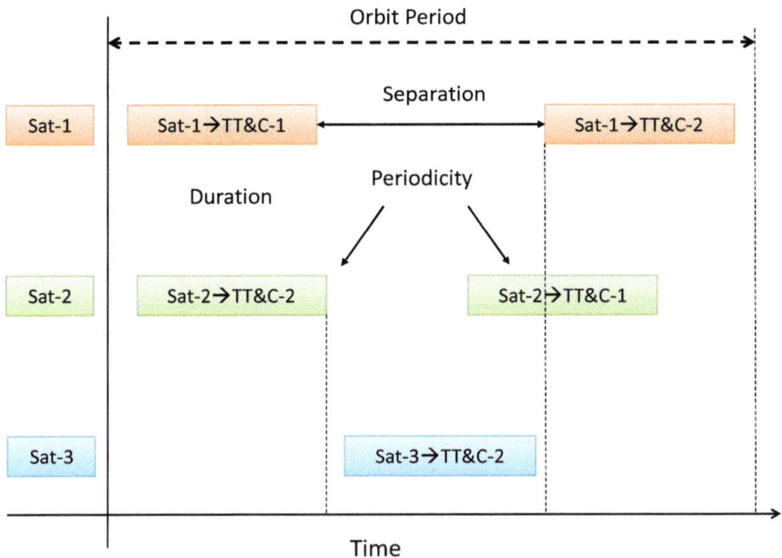

Fig. 8.6 Example of contact schedule for 3 satellites and 2 TT&C ground stations

- The contact separation time is defined as the elapsed time between the end of one contact and the start of the consecutive one to the same satellite. The maximum separation time could be driven by limitations of the onboard data storage capacity and the need to downlink data before it gets overwritten.
- The contact periodicity is defined as the minimum number of contacts that need to be scheduled for each satellite within one full orbit period. This parameter can depend on both satellite platform or payload specific requirements.

All of the aforementioned parameters are graphically depicted in Fig. 8.6 which gives an example for a simple contact schedule of three satellites (referred to as Sat-1, Sat-2, and Sat-3) and two TT&C ground stations. Even for this simple set up and the short window of only one orbit period, resource limitations evolve quite quickly. The contact schedule for the first two satellites only leaves one possible contact (per orbit) for the third one. This would imply a violation of a contact periodicity requirement, if a minimum of two contacts per orbit would be required. A possible conflict solving strategy in this case could be to either shorten the contact duration or relax the periodicity requirement for the third satellite.

8.5 Interfaces

An overview of the various interfaces of the MPF to other GCS elements is depicted in Fig. 8.7 and briefly described below:

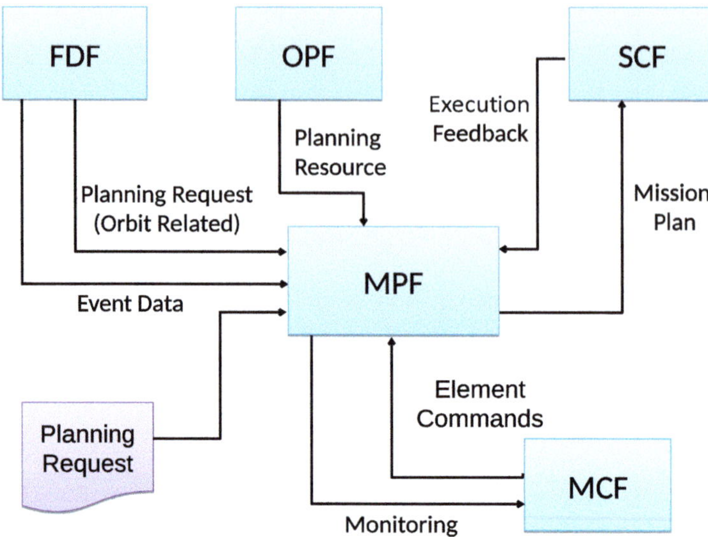

Fig. 8.7 Overview of the most important MPF interfaces to other GCS elements

- The **planning request** (PR) is the main driver of the planning process and needs to contain all the relevant information to allow the corresponding tasks and activities to be planned according to the initiator's needs or preferences. A meaningful PR needs to be clear on its scope and timing needs so the planning function can allocate the correct set of tasks and activities (goal based planning concept). As discussed in the context of conflict resolution, it is quite useful to provide margins on the requested start (or end) times in order to provide a certain amount of flexibility to the contact scheduler. The example of earliest, preferred, and latest start time has been mentioned earlier as a recommended rule (refer to Fig. 8.6). Another important characteristic of a PR is the definition of a PR life cycle that needs to be applied during the planning process. In line with the terminology introduced in CCSDS-529.0-G-1 [1], a PR can go through the following stages:

  - *Requested*: this is the initial state right after the PR has been submitted to the MPP.
  - *Accepted*: the PR has passed the first consistency check and can be considered in the MPP.
  - *Invalid*: the PR contains inconsistent information or does not follow the formatting rules and is therefore rejected. It needs to be corrected and resent for a subsequent validation.
  - *Rejected*: the PR has been considered in the MPP but could not be planned due to constraints or for any other reason. The possibility to provide information on the reason for the rejection should be considered in the definition of the interface.

- *Planned*: the PR could be successfully processed and has been added into the mission plan. Further processing will now depend on the reception of the execution feedback signal.
- *Executed*: the tasks and activities related to the PR have been executed and the execution status information is available.
- *Completed*: this is an additional evaluation of the execution status and might be relevant for PRs that depend on certain conditions that must be met during the execution. An example could be a scientific observation which might depend conditions that can only be evaluated, once the observation has been performed. This PR state is not meaningful for every project and might therefore not always be implemented.
- *Failed*: the scheduled tasks and activities related to the PR were not correctly executed. The reason could be an abort due to a satellite or ground segment anomaly.

The source of a PR can be a manual operator interaction by a mission planning engineer sitting in front of the MPF console or stem from an external source like the FDF as shown in Fig. 8.7. An example for an FDF originated PR is the request to perform an orbit correction manoeuvre or the uplink of new orbit vectors for the OOP.

- **FDF** (see Chap. 6): the event data file contains all satellite orbit related information and is based on to the most up-to-date satellite state vector information. The type of events might differ among various satellite projects depending on the specific planning needs. Typical event file entries are ground station visibilities, satellite eclipse entry and exit times, orbit geometry related events like node or apogee/perigee crossings, co-linearities with celestial bodies, or events related to bodies entering the Field of View (FOV) of satellite mounted sensors.
- **SCF** (see Chap. 7): the transfer of the Mission Plan (MP) to the Satellite Control Facility is the main output of the planning process and should contain a sequence of contacts with scheduled tasks and activities that accommodates all received PRs. The MP needs to be conflict free and respect all known planning constraints and contact rules and at the same time ensure an optimised use of all project resources. The format and detailed content of the MP will be very project specific. In the simplest case, it could be a mission human readable timeline or Sequence of Events (SOE) directly used by operators. In a more complex case, the MP could be realised as a file formatted for computers based reading and contain an extended list of contact times and detailed references to satellite procedures which can be directly ingested and processed in automated satellite operations. The execution feedback is a signal sent from the SCF to the MPF and contains important information on the execution status of a each task or activity and is based on the most recent available satellite telemetry. It allows to progress the life cycle state of the PR from planned to executed and finally to completed. Depending on the project needs, the execution status can either be sent to the MPF on a regular interval (e.g., after the completion of a task) or on request.

- **OPF** (see Chap. 9): planning resources comprise planning databases, constraints, and contact rules. A planning database should as a minimum comprise the detailed definition of all tasks and activities specific to a satellite project. As any change in the planning resources can have a strong impact on the contact schedule and mission plan, this information should be managed as configuration items and kept under configuration control via exchange with the Operation Preparations Facility. Having a clearly defined source of all planning related assumptions, the outcome of a planning process can be easily understood and reproduced.
- **MCF** (see Chap. 10): the interface to the Monitoring and Control Facility ensures the central monitoring of the MPF element hardware and processes and provides a basic remote commanding functionality like the change of element operational modes.

Reference

1. The Consultative Committee for Space Data Systems. (2018). *Mission planning and scheduling.* CCSDS 529.0-G-1. Cambridge: Green Book.

Chapter 9
The Operations Preparation Facility

The importance of a thorough validation process for all operational products and their proper management as configuration data has already been emphasised in the ground segment architectural overview in Chap. 4. The functionality of a segment wide configuration control mechanism for this data is realised in the Operations Preparation Facility (OPF) who's main functionalities can be summarised as follows:

- To import and archive all ground segment internal and external configuration items (CIs).
- To support the validation of CIs in format, contents, and consistency.
- To provide tools for the generation and modification of CIs (e.g., satellite TM/TC database, Flight Control Procedures, etc.).
- To provide means for a proper version and configuration control of all ground segment CIs.
- To distribute CIs to the correct end user for validation and use.
- To serve as a gateway between the operations and validation chains (OPE and VAL) in case a chain separation concept has been implemented (see Fig. 13.2).

9.1 Architectural Overview

A simplified architectural design of the OPF is depicted in Fig. 9.1. The core of this element is the CI archive which is used to store all segment CIs. Considering the vast amount of configuration files present on the various GCS elements, the overall number of CIs can be quite significant. In addition, the archive must also host the satellite specific CIs which have to be multiplied by the number of spacecraft, in case a satellite constellation is being operated.

© The Author(s), under exclusive license to Springer Nature Switzerland AG 2023
B. Nejad, *Introduction to Satellite Ground Segment Systems Engineering*, Space
Technology Library 41, https://doi.org/10.1007/978-3-031-15900-8_9

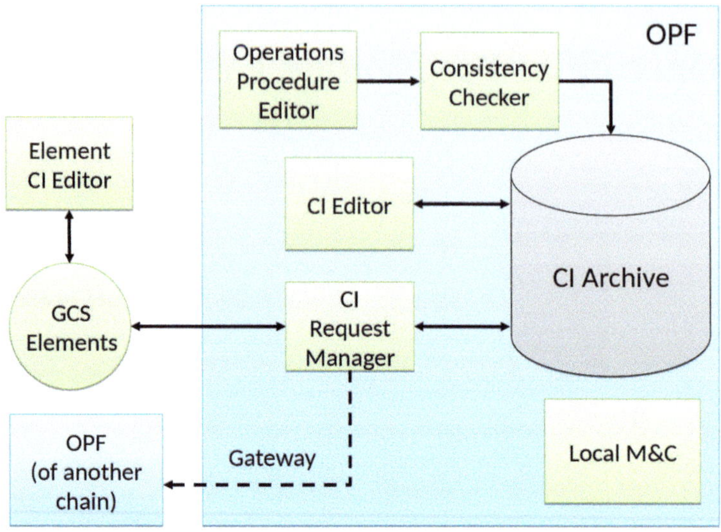

Fig. 9.1 Simplified architecture of the Operations Preparation Facility (OPF)

Configuration items can be grouped into different *classes* according to their content or their applicability. Satellite specific CIs can either apply to a single satellite *domain* or an entire *family*. The term family is frequently used if a constellation comprises spacecraft from different manufacturers which differ in their platform and sensor configuration. The CI archive structure might also consider the grouping of CI sets into different *series* based on their usage. One series could be dedicated for training purposes, one for software installation and validation, and one for the actual operations activities. Such a separation aims to avoid that any modification done during a training session or a software installation could alter a CI configuration that is used for actual flight operations.

The CI *editor* is a tool to modify the contents of a CI. For some CI types it might be more adequate to perform modifications at the element where the CI is operationally used. In this case, the element has its own CI editor and uses the OPF only as a means for configuration control and archiving. An example for such a CI modification at element level is shown in Fig. 9.2, where the element first sends a request to the OPF to reserve an existing CI version n. The CI Request manager receiving that request has to first identify, whether that CI exists in its archive and can then send a working copy back to the element. At the same time, the OPF prevents that any other user can request to modify that same CI. Once the CI editing at the element level has been finalised, the element operator needs to send the modified version back to the OPF. The OPF request manager then ensures that the updated CI is stored in the archive as a new version $n + 1$. The old version can still be retained (as version n), in case a roll-back to a previous configuration would ever be needed. The exact same principle for CI editing is also applicable when the

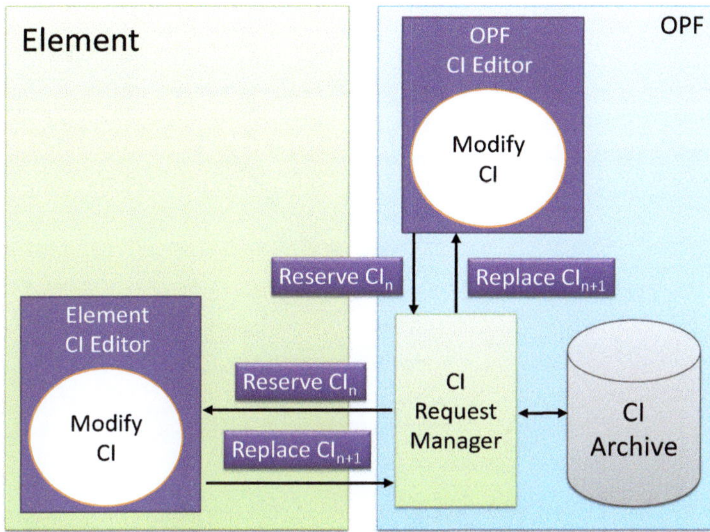

Fig. 9.2 Local element CI editing and interaction with OPF

OPF internal CI editor is used. The only difference is that the modifications would be done on the OPF element directly[1]

The *operations procedure editor* is a more advanced CI editor which is specialised to support the generation, modification, and validation of flight and ground operations procedures. These are usually referred to as FOPs and GOPs respectively and play an important role in satellite operations (see also description in the operations related Chap. 15). The *consistency checker* is a tool to verify that the TM/TC parameters which are referenced in the operations procedures are consistent with the applicable and most recent satellite databases (refer to SRDB definition in next section).

9.2 Configuration Items

The number of CIs, their contents, and categorisation is very project specific but some common types appear in most satellite projects which are described in more detail below.

---

[1] The CI modification policy (i.e., where the editing of a CI needs to be performed) should be decided for each CI class and driven by operational needs and aspects. It could also be the case that only one location for CI editing is sufficient which could simplify the GCS architecture in this respect.

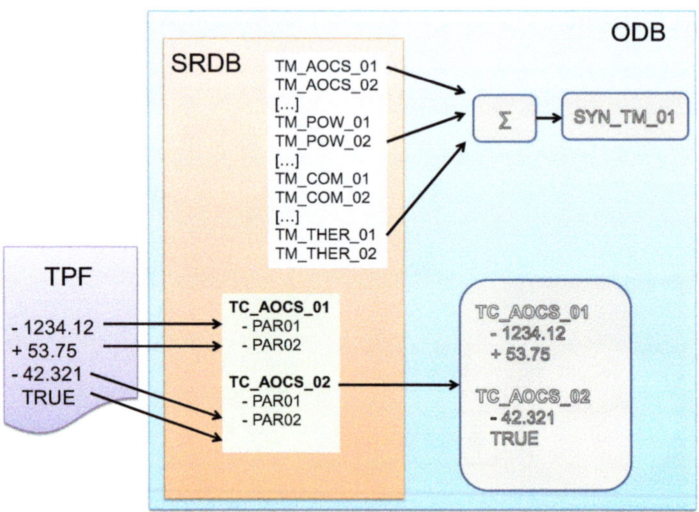

Fig. 9.3 Definition of derived parameters (synthetic TM) and Command Sequences with rules to populate TC parameters from external input files. TPF = Task Parameter File, SRDB = Satellite Reference Database, ODB = Operational Database

- The *satellite reference database* (SRDB) contains the full set of TM and TC parameters of a satellite. As the SRDB needs to be derived from the satellite on-board or avionic software, it is usually provided by the satellite manufacturer. The SRDB should therefore be considered as a contractual deliverable[2] from the space to the ground segment and needs to be tracked in terms of contents, delivery schedule, and delivery frequency. The SRDB might however not be the final product used for operations. It will usually be extended with additional derived (or synthetic) telemetry parameters which are generated by the database analyst, who combines the state or value of existing TM parameters to form a new one according to a defined logic or rule (see Fig. 9.3). This extended database is called the *operational database* (ODB) and can, in addition to the derived parameters, also contain definitions of telecommand sequences and rules needed to populate TCs with values from an external source. A good example would be the FDF providing AOCS related TC values (e.g., manoeuvre command parameters) via an input file to the SCF to populate the relevant TC sequences. The SCOS2000 standard (refer to [1]) provides a specific definition for such an interface file which is referred to as a *Task Parameter File* (TPF). It is important to note that the OPF has to be designed to manage both, the SRDB and the ODB for each satellite domain.

---

[2] Such products are usually referred to as a *Customer Furnished Item* or CFI.

- The *satellite flight dynamics database* (FDDB) contains all satellite specific data required by flight dynamics in order to generate satellite platform specific products. The content can be categorised into static parameters which remain constant throughout the satellite lifetime and dynamic data subject to change with certain satellite activities. Examples for static data are the satellite dry mass or the position of sensors and thrusters and their orientation (see also Table 6.1). An obvious dynamic parameter is the propellant mass which decreases after each thruster activity. Less obvious is the dynamic character of the satellite centre of mass location which moves within the satellite reference frame, whenever the propellant mass changes. The FDDB is a very important deliverable provided by the satellite manufacturer. Depending on the availability of intermediate and final versions, a staggered delivery approach should be agreed in order to gain an early visibility at ground segment level. A possible scenario would be to agree a delivery of *design* data (i.e., expected values) during the early satellite design and build process which is followed by updates on sensor and thruster position and orientation values obtained in the alignment campaign. A final delivery could be done after the satellite fuelling campaign has been completed which provides values of the wet mass and related updates of mass properties (e.g., inertia matrix).
- Flight and ground operations procedures (FOPs and GOPs) form another important set of CIs which were already briefly discussed in the context of the OPF CI editor. These procedures are usually written and maintained by the flight control team (FCT) and require several updates throughout the satellite lifetime due to their complexity. The OPF is an important tool for their development, archiving, and configuration control.
- The satellite on-board or avionics software (OBSW) is developed by the satellite manufacturer and needs to be delivered to the ground segment as a fully representative image ready for integration into the satellite simulator (SIM). The SIM serves as a representative environment for the validation of the operational procedures prior to their use in real operations and is explained in more detail in Chap. 11. The satellite manufacturer might need to develop software updates (patches) during the satellite lifetime and has to provide these to the ground segment. The OPF should serve as the formal reception point in order to guarantee the proper validation (e.g., completeness and correctness) and configuration control of the received products. A new OBSW image should always be distributed as a CI from the OPF to the simulator.
- The planning data CI comprises contact rules, conflict solving rules, and definitions of tasks and activities which are needed for mission planning (see also Chap. 8). This CI is a good example for a GCS internal CI, as it is generated and maintained without any segment external interaction. All the other examples given above can be categorised as external CIs.

9.3   OPE-VAL Gateway

Many ground segment architectures might implement a chain separation concept which isolates the validation environment (VAL chain) from the operational one (OPE chain). This is typically realised via a physical separation at network level to prohibit any direct data exchange (see also Fig. 13.2). As there is however a need to transfer validated CIs from the VAL to the OPE chain, the implementation of a secure transfer mechanism referred to as a *gateway* can be considered with a schematic concept shown in Fig. 9.4. To initiate a transfer, the OPF on the VAL chain sends the products not directly to the OPE chain but to a reserved directory on an intermediate gateway server which is referred to as the *VAL inbox*. The incoming data can now be scrutinised for viruses and malware using specialised software routines that are installed on the gateway server. Only if this check has been passed, the forwarding of "clean" data to a different area referred to as *OPE outbox* is performed. This location is visible for the OPF on the OPE chain which can initiate a data import. The deployment of additional firewalls and intrusion detection devices in between help to furthermore harden the security at network level.

A similar approach could also be implemented in the opposite direction (i.e., from OPE to VAL) but this transfer needs to be in line with the applicable project security requirements, keeping in mind that the OPE chain hosts satellite TM data that might be considered as sensitive information.

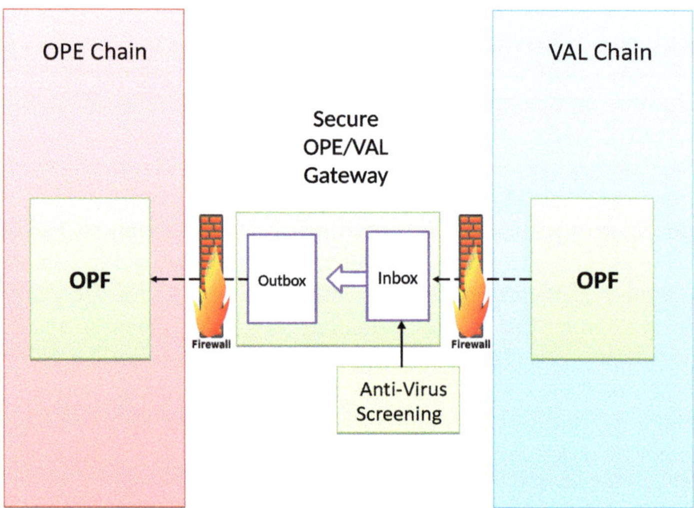

Fig. 9.4 Simplified architecture of the OPE/VAL Gateway

In general, any type of connection or transfer mechanism between the operational chain and any other environment is a highly sensitive topic and requires the consideration of highly secure safety measures.[3]

9.4   Interfaces

An overview of the various interfaces of the OPF to other GCS elements is depicted in Fig. 9.5. As can be readily seen, the OPF needs to be able to exchange CIs to all elements in the GCS. Even if the TT&C station is not explicitly shown here, it is in theory not excluded. The transfer mechanism needs to be designed to operate in both directions to support a CI reservation, update, and replacement protocol as described earlier in the text.

The interface to the Monitoring and Control Facility (see Chap. 10) is used to monitor the functionality of the OPF element and send commands. Example of commands could be to change element operational modes and even initiate CI exchange if considered as useful.

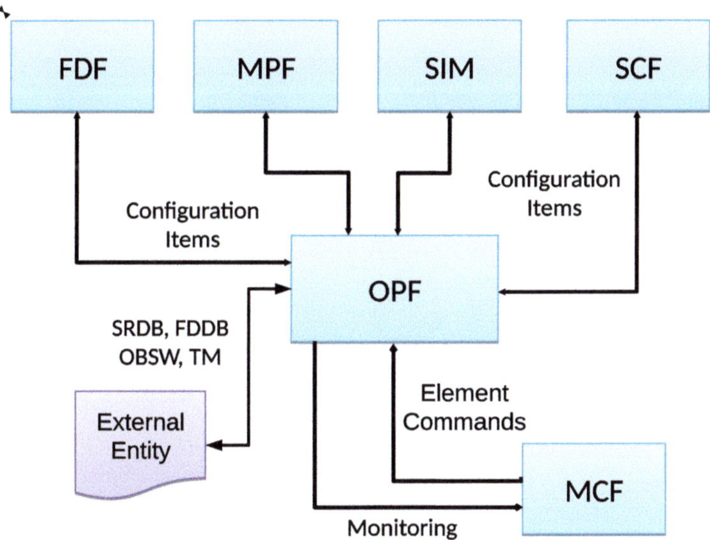

Fig. 9.5 Overview of main OPF interfaces

---

[3] Projects with high security standards put a lot of emphasis on the implementation of a highly secure network design that ensures no harmful data ingestion, transfer, export, or forbidden user access. Audits and penetration tests designed to proof the robustness of the architecture might form part of an accreditation process which can even be considered as a prerequisite to grant the *authorisation to operate.*

The implementation of an interface between OPF and an external entity is driven by the need to exchange satellite design specific CIs with the satellite manufacturer (e.g., SRDB, onboard software images or patches, FDDB, etc.). Being an external interface, the design needs to consider additional security requirements and should consider a bidirectional flow to allow operational telemetry to be shared with the satellite manufacturer during the early project phases (e.g., LEOP or in-orbit testing), in case support for potential failure investigations needs to be provided.

Reference

1. European Space Agency. (2007). *SCOS-2000 Task Parameter File Interface Control Document, EGOS-MCS-S2K-ICD-0003*.

Chapter 10
The Monitoring and Control Facility

The functional readiness of the ground segment depends on the proper functioning of all its components which can comprise a significant number of servers, workstations, network devices (e.g., switches, routers, or firewalls), and software processes. The Monitoring and Control Facility (MCF) is responsible to provide such a functional overview and inform the operator in case an anomaly occurs. In more detail, the following minimum set of functional requirements need to be fulfilled:

- The collection of monitoring data at software and hardware level for all major elements and components in order to provide a thorough, reliable, and up-to-date overview of the operational status of the entire ground control segment. Special focus should be given to any device that is essential to keep the satellite alive.
- The ability to detect and report an anomaly and report it with minimum latency to the operator to ensure that informed actions can be taken in time to avoid or minimise any damage. The type, criticality, and method of reporting (e.g., visual or audible alerts) should be optimised to ensure an efficient and ergonomic machine operator interaction.
- To provide the user the ability to command and operate other ground segment elements. Emphasis should be given to infrastructure which is located remotely and might note be easily accessible (e.g., an unmanned remote TT&C station).
- The ability to generate log files that store all the monitored data in an efficient format which minimises the file size but does not compromise on data quality. The file format should allow the use of a replay tool which eases any anomaly investigation at a later stage.
- To provide the means for remote software maintenance for elements that are not easily accessible.

© The Author(s), under exclusive license to Springer Nature Switzerland AG 2023

B. Nejad, *Introduction to Satellite Ground Segment Systems Engineering*, Space Technology Library 41, https://doi.org/10.1007/978-3-031-15900-8_10

10.1 Architectural Overview

A high level MCF architecture addressing the aforementioned functional requirements is depicted in Fig. 10.1. The presented architecture should only be viewed as an example, as the detailed design might again be very project specific and will strongly depend on the overall ground segment design and complexity.

The central component in Fig. 10.1 is the *online M&C manager* which is in charge of the reception and correct processing of all the incoming monitoring data from the various elements and their components. The measuring, formatting, and transmission of monitoring data is performed by the *host agents* which are software components installed on the servers of the monitored entities and have access and permission to inquire information about all the hardware and relevant software processes. The host agents can run as a background process[1] in order to avoid any disturbance with the main element activity. Careful consideration has to be given to find an appropriate polling rate for each monitoring parameter, as the need will differ depending on the type of hardware or software process under scrutiny. A reasonable starting point to estimate the right polling rate is to analyse how fast a monitoring parameter changes over time. One should also keep in mind that a high polling rate increases the data volume that needs to be transmitted to the central M&C manager which puts a higher load on the used network bandwidth, CPU processing

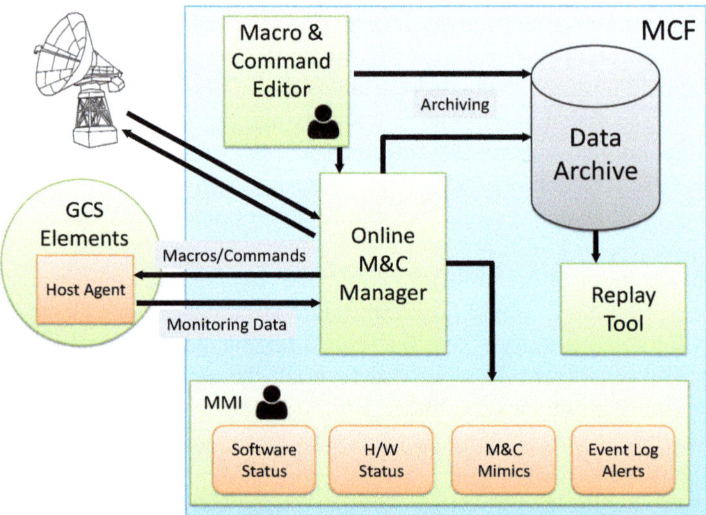

Fig. 10.1 High-level overview of the Monitoring and Control Facility (MCF) architecture

---

[1] Most operating systems allow the execution of software *daemons* which can be automatically started when a computer system is booted.

power, and required disk space for archiving. A polling rate that has been set too low might however increase the risk to loose the detection capability for highly dynamic parameters.

Once all the required post-processing is done, the M&C manager must forward the monitoring data to the Man Machine Interface (MMI) component which is in charge for data presentation to the operator. Depending on the ground segment complexity, there might be a considerable amount of monitoring data and events to be displayed. A good and efficient MMI architecture is therefore a crucial part of the overall MCF design and some relevant aspects for consideration are therefore discussed in Sect. 10.3.

The ability to remotely operate other GCS elements is another important task of the MCF. For satellite projects which require the use of unmanned TT&C ground stations which are deployed at remote locations distributed around the globe, the MCF can be the main access point for their daily operation. In this context the *macro & command editor* is an important tool to develop, modify, and potentially even validate command sequences which are sent to the GCS elements through the M&C manager.

All the monitoring data received by the MCF must be stored in the *data archive*. This archive needs to be properly dimensioned in terms of disk space in order to fulfil the project specific needs. The disk size analysis needs to consider the actual polling rates and compression factors that are foreseen for the various types of monitored parameters. A tool for the replay of the monitoring data is not mandatory but is a convenient means to investigate anomalies observed in the ground control segment.

10.2   The Simple Network Management Protocol

The remote management[2] of devices being connected via a network requires the implementation of a communication protocol. As the specification (and validation) of such a protocol requires a considerable amount of effort, it is more efficient to use an existing one that is standardised and has already been extensively used.

A very prominent remote network management protocol that meets these requirements is the *Simple Network Management Protocol* or SNMP. It has been published in its first version in 1988 [1], soon after the TCP/IP internet protocol was established and widely used which generated a strong demand to remotely manage network devices like cable modems, routers, or switches. The SNMP protocol has since then undergone three major revisions, mainly to implement improvements both in flexibility and security. The overall interface design is based on the following three building blocks (see also Fig. 10.2):

---

[2] Remote management in this context refers to the ability to collect monitoring data of a remote device and to send commands to modify its operational state or configuration.

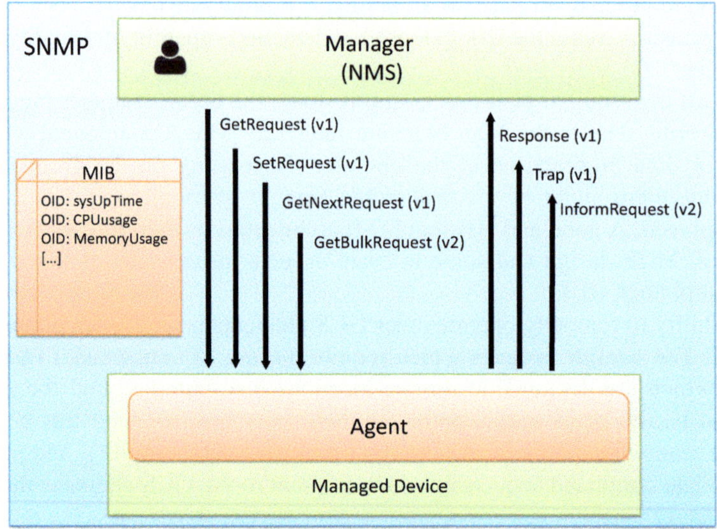

Fig. 10.2 The Simple Network Management Protocol (SNMP) architecture and messages. MIB = Management Information Base, NMS = Network Management System, OID = Object Identifier

- The SNMP *manager* (also referred to as network management station or NMS) is a software that is installed on the machine which is intends to monitor and control other devices. In the earlier presented MCF architecture, it is deployed in the online M&C manager.
- The SNMP *agent* runs on the device which needs to be monitored. In the case of the MCF, it corresponds to the earlier described host agents.
- The *Management Information Base* or MIB file defines the contents (scope) and structure of monitoring data. In other words, it defines what exactly the *managing* device can request from the *managed* device. Each variable specified in the MIB is identified by its unique *object identifier* or OID which is used to either access the variable's contents or to modify it according to the SNMP protocol rules.

The information exchange between the manager and the agent is organised according to SNMP protocol data unit (PDU) types. In the current version (i.e., SNMPv3), there are seven PDUs of which four go from the manager to the agent and the remaining three into the opposite direction (see arrows in Fig. 10.2). The *GetRequest* and *SetRequest* types are used to retrieve and change the value of one variable. The *GetNextRequest* queries the value of the subsequent object in the MIB and is used in an iterative loop to query the full set of objects. The *GetBulkRequest* has only been added in SNMPv2 and can be considered as an optimised version performing multiple iterations of the *GetNextRequest*. The *response* type allows the agent to provide a response on various request types. The *trap* is of special importance as it allows the host to notify the manager. As the transmission of a trap does not require any a preceding management request it enables the host to notify the manager of significant events and the *InforRequest* supports the trap allowing the manager to provide an acknowledgement after having received a trap.

10.3 Man Machine Interface

The Man Machine Interface (MMI) of the MCF must provide an efficient and ergonomic design that allows an operator to gain a quick and accurate overview on the overall GCS status both at software and hardware level. In case of problems, the MMI should provide easy access to a more detailed set of information and be able to display additional monitoring data at element and component level. The most important MMI tasks are to report a non-nominal behaviour as soon as possible (preferably via visible and audible means) and to subsequently support the impact analysis and recovery process through the provision of accurate information on the error source.

An example for a MMI design is shown in Fig. 10.3 which, despite its simplicity, shows the most relevant elements one can also find in complex systems. The area on the upper left provides a number of buttons that allow the operator to switch between different views. The currently active one (shown in a different colour) provides the

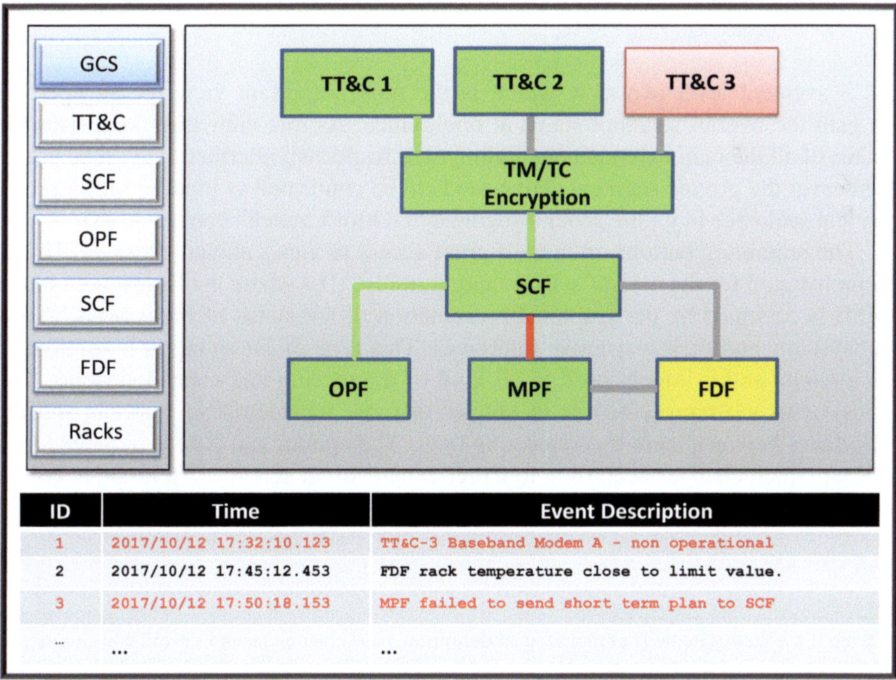

ID	Time	Event Description
1	2017/10/12 17:32:10.123	TT&C-3 Baseband Modem A - non operational
2	2017/10/12 17:45:12.453	FDF rack temperature close to limit value.
3	2017/10/12 17:50:18.153	MPF failed to send short term plan to SCF
...	...	...

Fig. 10.3 Example of a possible MCF MMI for a highly simplified GCS

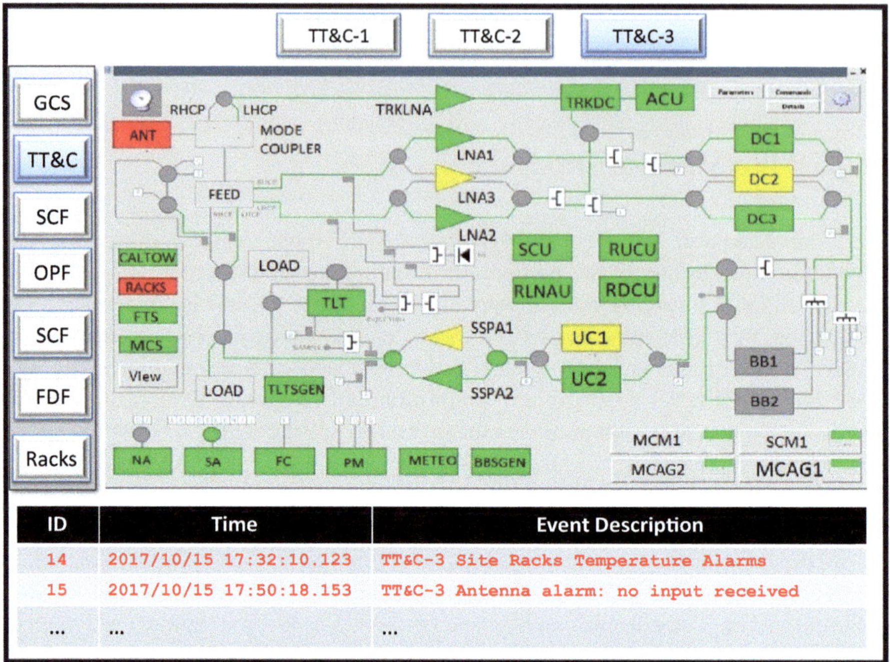

ID	Time	Event Description
14	2017/10/15 17:32:10.123	TT&C-3 Site Racks Temperature Alarms
15	2017/10/15 17:50:18.153	TT&C-3 Antenna alarm: no input received
...	...	...

Fig. 10.4 The TT&C mimic as an example for a mimic at element level

GCS segment level overview which is the most important view for an operator to gain the overall segment status at one glance. As this view also comprises the status of all the major elements including their respective interfaces, it will be useful whenever the ground segment status needs to be confirmed as input to the start of a critical sequence (e.g., the ground segment "GO for Launch" confirmation).

The remaining buttons on the left grant access to views at element level. This is demonstrated for the case of a TT& station in Fig. 10.4 where the largest area of the MMI is occupied by the graphical presentation of the status of the various TT&C subsystems and their respective interfaces. This type of presentation is referred to as a *mimic* and is widely used in all kind of monitoring and control systems. The subsystems are represented as rectangles or more representative symbols and the interfaces between them by connecting lines. The operational state is indicated by a colour which changes according to a predefined colour scheme. An intuitive colour scheme could be defined as follows:[3]

---

[3] Even if a colour scheme is pure matter of definition, care must be taken to avoid non intuitive or highly complex ones. As an example, the choice of the colour green to mark a critical event will very likely be misinterpreted. A scheme with too many colours, e.g., pink, violet, brown, or black, might not be obvious and requires a clear definition to be correctly interpreted.

- Green: the subsystem is in a nominal state and no operator intervention is needed.
- Yellow: the subsystem is not nominal but has not reached a critical state yet. For a redundant subsystem, this could mean that it is in standby mode and waiting to be activated.
- Red: the subsystem is reporting parameters in a critical (outside nominal range) state. The operator needs to act immediately and start the investigation. A fast recovery must be high priority in order to avoid the loss of ground segment operability.
- Gray: the subsystem does not report any monitoring data. Either it is disconnected, switched off, or there is a loss of connection. Whether this is critical or nominal, depends on the architecture and expected state of this subsystem.

The level of detail (e.g., number of elements, subsystems, components, parts) presented in a mimic depends on the complexity of the monitored system and the level of monitoring that is actually necessary (or meaningful). A well designed mimic must therefore undergo a thorough analysis in order to achieve a balanced trade off between detail and good overview. A higher level of detail will on one hand lower the risk to miss important events but on the other hand contribute to the "loading" of the visual presentation which increases the risk for an operator to overlook an important event (human error factor).

The lower parts of the MMIs in Figs. 10.3 and 10.4 are occupied by the event logger that presents a list of events in tabular and chronological order. It must be clearly defined whether the reported time refers to the actual time of occurrence or the time of reporting, which could be different due to the polling rate and transmission latency. A colour code to distinguish the criticality of events is also recommended for the logger. Furthermore, the use of audible alarms for highly critical events can further improve the operator notification.

From an operational point of view, the MCF user would proceed from the higher level to the lower level views in order to investigate the type and source of an anomaly. In the example view in Fig. 10.3, the operator starts from the segment (GCS) view which indicates various problems. The TT&C-3 station is shown in red indicating that it is not operational anymore. The segment log states a problem with the TT&C-3 baseband modem which could be one reason for the station unavailability. To gain further insight, the operator must switch to the TT&C specific view where the two (redundant) baseband modems "BB1" and "BB2" are shown in grey colour pointing to a loss of monitoring. This is clearly not an expected state for such a critical subsystem. Furthermore, both the antenna subsystem and site rack boxes are in red and corresponding messages are shown in the TT&C log window.

Going back to the segment level view, an interface anomaly between the MPF and the SCF is indicated by the red colour of the connecting line which has prevented the exchange of planning data between these two elements (see also corresponding log message concerning the short term plan). Finally, the FDF box is shown in yellow which indicates a warning of a potential hardware problem as the rack temperature is elevated and close to reach the allowed upper limit. Even if no immediate action is needed, the issue requires close monitoring and investigation by a local team having access to that hardware.

The example above clearly demonstrates the ability of simple but well designed MMI to provide a quick overview of a number of anomalies in a complex infrastructure. The log also plays an important role as a short and simple text based message can already provide a good indication of the time, location, nature, and criticality of an anomaly which allows to narrow down the root cause investigation.

10.4   Event Replay & Archiving

In a realistic operational scenario there will be thousands of events displayed in the MCF event log every day. It is therefore quite important that the event logger implements a regular archiving and backup policy which allows this huge amount of information flow to be scrutinised at a later stage. The archiving of the mimic views in connection with the log messages further eases an anomaly investigation. The use of modern file compression techniques should allow to keep the data footprint small without significant loss of information. The integration of a replay tool which is compatible with the file format allows to easily reconstruct any historic state of the entire ground segment and its subsystems.

10.5   Virtual Machine Monitoring

A typical virtualisation architecture usually comprises several physical servers with each of them implementing a type-1 hypervisor that hosts a number of virtual machines (VMs). The fundamental set-up of such a virtualised system is described in Chap. 14 and shown in Fig. 10.5.[4] Most virtualisation solutions currently available on the market provide their own monitoring and control application (as a COTS) which is referred to as a *virtual management centre* (VMC)[5] (see upper box in Fig. 10.5). The VMC can be used for the management (e.g., creation and configuration) and monitoring of all VMs and their physical hosts and is able to provide standard hardware monitoring data like the use of CPU power, memory, or disk space and also monitoring data of active software processes.

From a ground segment architecture point of view, the VMC application can be either integrated into the overall MCF design and would then add its VM based monitoring data to the overall monitoring data pool, or be used as a (standalone) monitoring system.

---

[4] The intention of this section is to make the reader aware of the additional monitoring and control functionalities of a virtualised system architecture.

[5] The actual name of the VMC application will very much depend on the virtualisation software being used, one example is VMware's vCenter Server[®] [2].

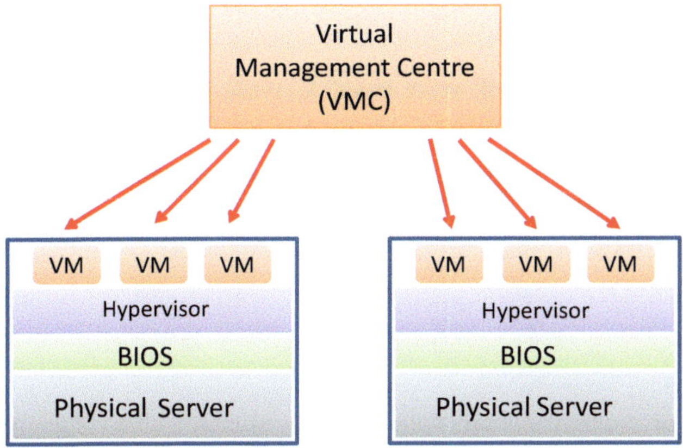

Fig. 10.5 VM and physical hosts monitoring by the Virtual Management Centre

Fig. 10.6 Overview of main
MCF interfaces

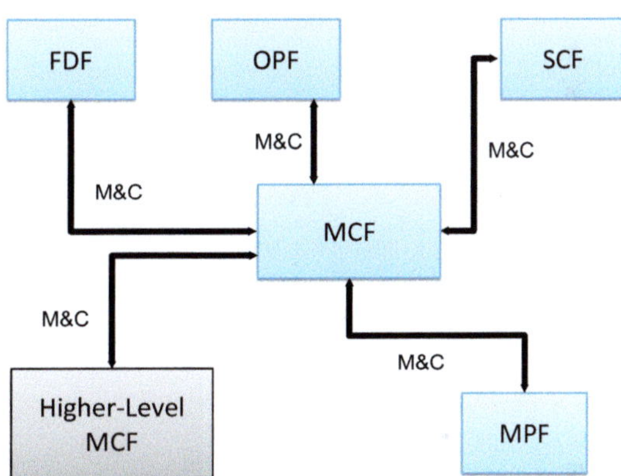

10.6 Interfaces

An overview of the various interfaces of the MCF to other GCS elements is
shown in Fig. 10.6. As can be readily seen, the MCF requires the implementation
of a dedicated (snmp) interface to each GCS element and the flow direction is
always from element to MCF for monitoring data and the opposite direction for
commanding. An additional interface to a higher level MCF is also shown which is
however only needed if a staggered monitoring and control architecture is realised.
This could be the case if the GCS itself is only a subsystem of a larger ground
segment architecture and an additional MCS is operated at a higher hierarchical
level. In this case, the GCS MCF could filter and correlate all the GCS level
monitoring data and forward these to the higher level MCF.

References

1. Case, J., Fedor, M., Schoffstall, M., & Davin, J. (1988). Network working group request for comments 1067: A simple network management protocol. https://www.rfc-editor.org/rfc/pdfrfc/rfc1067.txt.pdf Accessed Mar 14, 2022.
2. Vmware Vcenter Website. (2022). https://www.vmware.com/de/products/vcenter-server.html Accessed Mar 14, 2022

Chapter 11
The Satellite Simulator

In a space project, every mistake can have a serious impact on the satellite and its ability to provide its service to the end user. In a worst case scenario, an incorrect telecommand could even lead to a permanent damage of flight hardware or even the loss of a satellite. It is therefore worth to scrutinise the main sources for mistakes in flight operation which will typically fall into one of the following categories:

1. The specification of incorrect or inaccurate steps in flight operations procedures (FOPs).
2. The incorrect execution of a (correctly specified) step in a FOP due to a human (operator) error.
3. An incorrect behaviour of a GCS element or an interface due to a software bug or hardware malfunctioning.

A representative satellite simulator can reduce the risk of any of these scenarios to occur by satisfying the following functional requirements:

- To support the qualification (i.e., verification and validation) test campaigns of the GCS and its elements during their development, initial deployment, and upgrade phases.[1] The Satellite Control Facility (SCF) has a direct link the simulator and will therefore be the main beneficiary. It can send TCs to the simulator and receive TM that is representative of the actual satellite onboard software, sensors, and actuators and their behaviour in space.
- To provide a test and "dry-run" environment for the validation of an onboard software patch prior to its uplink to the real satellite.
- To provide a training environment to the flight control team (FCT) that allows routine and emergency situations to be simulated in a realistic manner. Even if the risk factor of human error can never be entirely excluded, the likelihood of

---

[1] In this context the simulator is also referred to as the Assembly Integration and Verification (AIV) Simulator.

© The Author(s), under exclusive license to Springer Nature Switzerland AG 2023
B. Nejad, *Introduction to Satellite Ground Segment Systems Engineering*, Space Technology Library 41, https://doi.org/10.1007/978-3-031-15900-8_11

occurrence can be significantly reduced through the appropriate training of all FCT members. Appropriate in this context has to address the training duration, level of detail, and the level of representativeness of a real operational scenario, and the simulator can clearly contribute to the last one of these aspect.

- To provide a validation environment for flight and ground operations procedures (FOPs and GOPs) prior to their use in real operations. In case of an onboard software update, bot FOPs and GOPs might have to be adapted and undergo a revalidation campaign.

To properly address these needs, the simulator design needs to incorporate the following aspects.

- The ability to accurately emulate the satellite onboard software and its interface to the ground segment.
- An accurate modelling of sensors and actuators which provide a (flight) representative sensor feedback to the onboard software.
- The correct and accurate emulation of all satellite system and attitude modes as well as their possible transitions.
- A representative simulation of the space to ground segment link, considering the correct performance parameters for the bandwidth and latency.

11.1   Architectural Overview

An example of a satellite simulator architecture with its major components is shown in Fig. 11.1. The *simulator environment* comprises all the non satellite specific components which support the run-time environment (kernel) that control the state of the simulator (e.g., transition between various simulation modes) and support the scheduling and command processing functions. Furthermore, the Man-Machine-Interface (MMI) is implemented in this part.

The *models* component contains all the satellite specific and space environment models. The satellite portion has to integrate a fully representative copy of the actual satellite onboard software (OBSW),[2] including its satellite reference database (SRDB) with the full list of TC and TMs definitions. Furthermore, models of all relevant satellite subsystems like AOCS sensors and actuators, data handling bus, transponders, etc. need to be realised, in order to allow a representative feed back to the OBSW.

The *environment* module provides the relative position of celestial bodies (e.g., Sun or Moon) with respect to the satellite. For this, the spacecraft orbit dynamics (position, velocity, and attitude) needs to be computed with a numerical state propagator that implements a realistic force model. In addition, the apparent size and luminosity of celestial bodies[3] can be relevant to correctly simulate their crossing

---

[2] Note that the terms onboard software and avionics software have similar meaning.

[3] For Earth bound satellites the Sun, Moon, and Earth will most likely be sufficient.

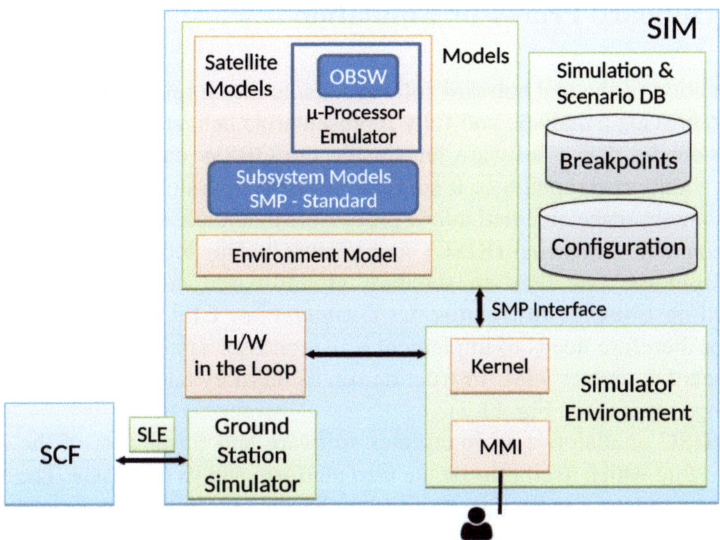

Fig. 11.1 High-level simulator architecture: SMP = System Model Portability (refer to ECSS-E-TM-40-07 [1]), SLE = Space Link Extension (refer to CCSDS-910-0-Y-2 [2])

of a sensor field-of-view. It is important to distinguish between the orbit dynamics computed by the environment module and the one from the OBSW. The latter one is based on the satellite internal *onboard orbit propagator* (OOP) which is also shown in the TM. Like in reality, the accuracy of the orbit derived from the simulator OOP orbit will depend on the time of the most recent OOP update.

The *simulation & scenario database* hosts all the satellite specific configuration files and simulation states at specific times during the satellite lifetime which are referred to as simulation *breakpoints*. One such breakpoint could represent the state just after the satellite separation from the launcher which is frequently used for LEOP simulation campaigns. Certain satellite configurations during the routine phase might also be useful breakpoints to simulate the transition into a specific slew configuration prior to the execution of an orbit control manoeuvre.

The *ground station simulator* component simulates the link between the TT&C station and the SCF based on the space communication protocol that is applicable for the satellite to ground interface. A widely used protocol in this context is the Space Link Extension (SLE) as defined by the CCSDS-913.1-B-1 [3] standard.

In case a satellite subsystem cannot be modelled via software, the simulator needs to integrate flight hardware which is shown as *H/W in the loop* in Fig. 11.1. An example for such a limitation could be the presence of a security module with a classified encryption algorithm that cannot be exposed to the outside world.

11.2   Onboard Processor Emulation

The execution of the real onboard software inside the simulator is the most efficient
means to provide a realistic and fully representative behaviour of the satellite. This
can also ensure that a software bug inside the OBSW can be discovered during
a FOP validation campaign or a simulation exercise. Satellite onboard computers
usually contain space qualified micro-processors that implement a so called *Reduced
Instruction Set Computer* (RISC) architecture.[4] The RISC architecture clearly
differs from the one used in standard ground-based computer systems which
are based on *Complex Instruction Set Computers* or CISC micro-processors. The
simulator therefore needs to implement a μ-processor emulator to properly handle
the on-board computer RISC instruction set, its memory addressing scheme, and all
the I/O operations (see Fig. 11.2).

The RISC emulator is an interpreter software and forms part of the simulator
software suite which itself runs on the *host* platform and its processor. The execution
of such an emulator can be quite demanding for the host, as for a real time emulation
of the *target* processor the host processor requires a 25–50 times higher performance

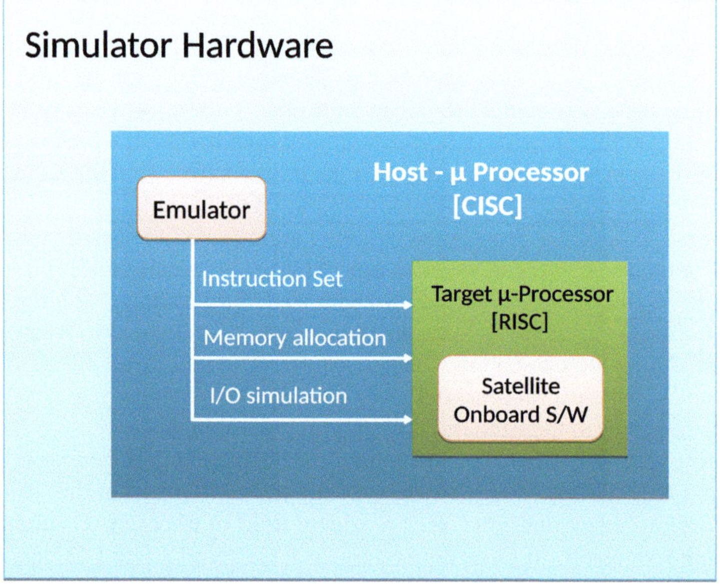

Fig. 11.2 The μ-processor emulator of the simulator: the *host* is the computer platform on which
the emulator is running (i.e., the simulator's processor) and the *target* is the emulated computer
hardware (i.e., the satellite onboard computer)

---

[4] Many flying ESA missions use for example the ERC32 and LEON2 processors which are based
on the SPARCv7 and SPARCv8 architecture [4].

(refer to [4]). This is even a conservative estimation, as it only applies for single core type target processors. For newer high performance quad-core processors (e.g., the GR740 radiation hardened multi core architecture processor developed under ESA's Next Generation Microprocessor (NGMP) program [5]), this emulation load is even more demanding and needs to be considered for the specification of the simulator hardware performance.

11.3 System Modelling

The development effort for a new satellite simulator can be reduced if existing simulation environments and models can be reused (ported) across platforms and projects. This is facilitated with the application of model based software development techniques and the use of model standardisation like the *Simulation Model Portability Standard* (SMP) proposed by ESA [6]. SMP can be used to develop a *virtual system model* (VSM) of a satellite[5] which must be understood as a meta-model that describes the semantics, rules, and constraints of a model and can therefore be specified as a *platform independent model* or PIM that does not build on any specific programming language, operating system, file format or database (see Fig. 11.3). If the PIM has been based on the rules and guidelines of the SMP standard[6] as suggested in ECSS-E-TM-40-07 [1], its translation into a *platform specific model* (PSM) is more easily achieved. According to the SMP specification as defined in the SMP Handbook [7], the PIM can be defined using the following three types of configuration files:

- The *catalogue file* containing the model specification,
- the *assembly file* describing the integration and configuration of a model instance, and
- the *schedule file* defining the rules for the scheduling of a model instance.

with the *Simulation Model Definition Language* (SMDL) defining their format as XML. The SMP specification also provides standardised interfaces for the integration with a simulation environment and its services. This is schematically depicted in Fig. 11.4 where a number of satellite models referred to as *Model 1* to *Model N* are able to connect to the simulation environments that are specific to two different projects *A* and *B* which both implement the SMP simulation services *Logger, Scheduler, Time Keeper, Event Manager, Link Registry*, and *Resolver*. The SMP standard also foresees interfaces for an inter model communication which furthermore facilitates the reuse of models among different projects and platforms.

---

[5] In ESA standards this satellite VSM is also referred to as the *Spacecraft Simulator Reference Architecture* or REFA.

[6] The most recent version today is referred to as SMP version 2 or simply SMP-2.

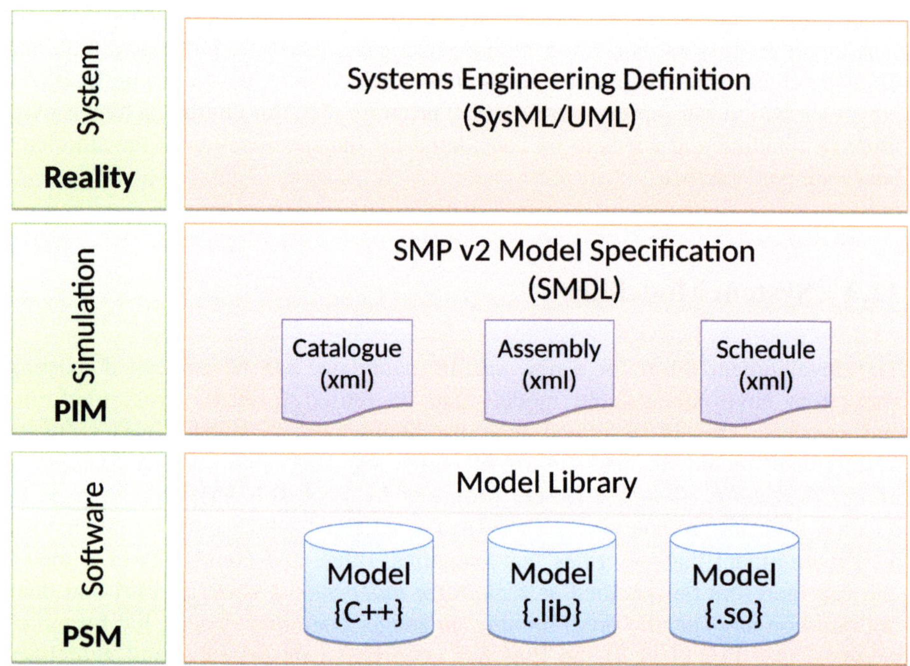

Fig. 11.3 Simulation Model Portability (SMP) development concept; SysML = System Modelling Language, UML = Unified Modelling Language, SMDL = System Modelling Definition Language, PIM = Platform Independent Model, PSM = Platform Specific Model

Fig. 11.4 Integration of SMP compatible simulation models to project specific simulation environments using SMP interfaces

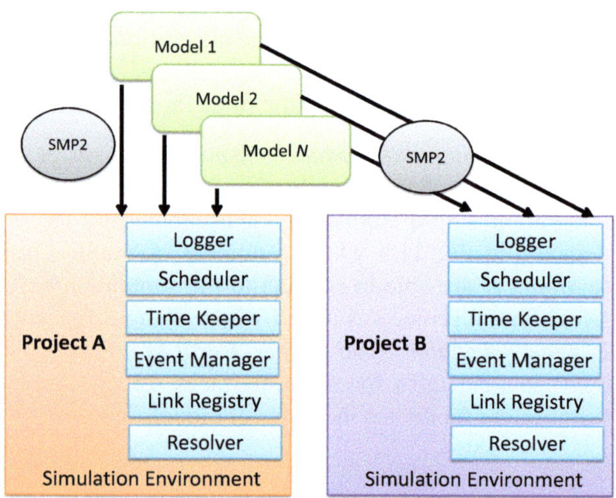

11.4 Interfaces

The most relevant interfaces between the ground segment simulator and the other GCS elements are shown in Fig. 11.5 and described in the bullets below:

- **FDF** (see Chap. 6): the interface to the Flight Dynamics Facility (FDF) is useful for the provision of either initial satellite orbit state vectors or even complete orbit files. In the latter case the simulator internal orbit propagation would be replaced by interpolation of the provided FDF orbit files. As historic orbit files generated by the FDF are based on orbit determination providing a very accurate orbit knowledge, their use can be highly beneficial in case a certain satellite constellation needs to be accurately reproduced in the context of an anomaly investigation. Furthermore, the exchange of space or orbit events (e.g., ascending node, perigee, apogee crossings, or sensor events) might be useful information for the cross correlation of events seen in the simulation.
- **SCF** (see Chap. 7): the most important interface is the one to the Satellite Control Facility for the exchange of TM and TC. Depending on the need to simulate classified satellite equipment with encrypted data, a separate data flow channel between the simulator and the SCF has to be implemented. This specific flow has to pass through the specific security units in each of the two elements which host the keys and algorithms to handle encrypted TM and TC (indicated by the red boxes and arrow in Fig. 11.5). Depending on the project needs and security rules, such encryption and decryption units might have to be housed in dedicated areas which fulfil the required access restrictions of a secure area.
- **MPF** (see Chap. 8): the implementation of a dedicated interface between the Mission Planning Facility (MPF) and the simulator can be useful to exchange

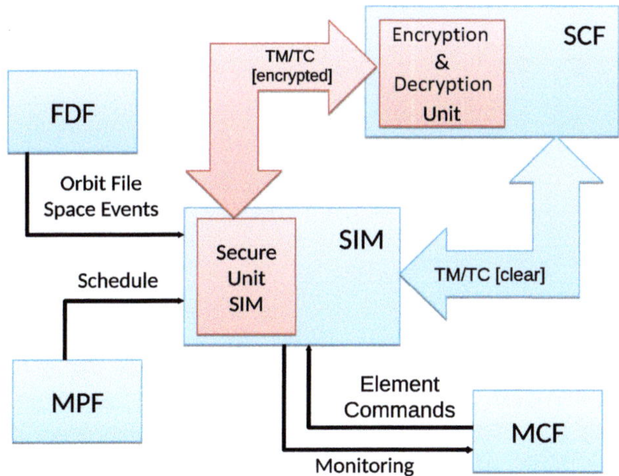

Fig. 11.5 Simulator interfaces with other GCS elements. Note that interfaces to FDF and MPF might not be realised in every ground segment

schedule relevant information. A typical example could be the down time of a TT& C ground station which might be impact the satellite contact times that are relevant for the simulation.

- **MCF** (see Chap. 10): the interface to the Monitoring and Control Facility (MCF) is less critical for the simulator than for other elements as this element is not directly involved in real time satellite operations. If a chain separation concept is part of the GCS architecture (refer to Sect. 4.2 and Fig. 13.2), the simulator would only be deployed on the validation chain. The implementation of this interface should however still be considered.

References

1. European Cooperation for Space Standardization. (2011). *Simulation modelling platform: Principles and requirements* (Vol. 1A). ECSS. ECSS-E-TM-40-07.
2. The Consultative Committee for Space Data Systems. (2016). *Space link extension services: Executive summary*. CCSDS. CCSDS 910.0-Y-2.
3. The Consultative Committee for Space Data Systems. (2008). *Space link extension: Internet protocol for transfer services*. CCSDS. CCSDS 913.1-B-1, Blue Book.
4. Holm, M. (2016). *High performance microprocessor emulation for software validation facilities and operational simulators*. AIAA. AIAA 2016-2643.
5. Hjorth, M., et al. (2015). GR740: Rad-hard quad-core LEON4FT system-on-chip. In *Programme and abstracts book of the DASIA 2015 conference, Spain.*
6. Walsh, A., Pecchioli, M., Reggestad, V., & Ellsiepen, P. (2014) ESA's model based approach for the development of operational spacecraft simulators. In *SpaceOps 2014 conference, 5–9 May 2014, Pasadena CA*. AIAA. https://doi.org/10.2514/6.2014-1866.
7. European Space Operations Centre. (2005). *SMP 2.0 handbook* (Issue 1, Rev. 2). European Space Operations Centre. EGOS-SIM-GEN-TN-0099.

Chapter 12
Auxiliary Services

Auxiliary services refer to all those tools and applications that are required for the proper functioning of a ground segment, but do not actually belong to any of the previously described functions or elements. These functions usually support the (centralised) management of user accounts, the data backup policy, an efficient and traceable transfer of data, but also security related aspects like the antivirus protection or the management of software releases and their version control. This chapter provides a description of these auxiliary services which should be considered in every initial ground segment design.

12.1 Centralised User Management

User account management is an important function in every multi user system and even more in a security sensitive facility like a satellite ground segment where a controlled and accountable user access must be part of the applicable security policy. User account management should provide the means to efficiently create, maintain, restrict, and if necessary remove any user access. In a ground segment and operations context the *role-based access control* or RBAC scheme shown in Fig. 12.1 is very suitable and should therefore be the preferred choice. The RBAC principle is based on the definition of a set of *roles* which are authorised to perform a set of *transactions* on objects, where this transaction can be seen as a transformation procedure and/or data access [1]. Once all the roles and their allowed transactions are well defined (and usually remain constant), the user administrator's task is to simply grant a new user the adequate membership to one or more specific roles, which will then automatically define the amount and type of resources the user can access. This scheme provides quite some flexibility as it allows one user to be member of several roles or even one role to be composed of other roles (see

B. Nejad, *Introduction to Satellite Ground Segment Systems Engineering*, Space Technology Library 41, https://doi.org/10.1007/978-3-031-15900-8_12

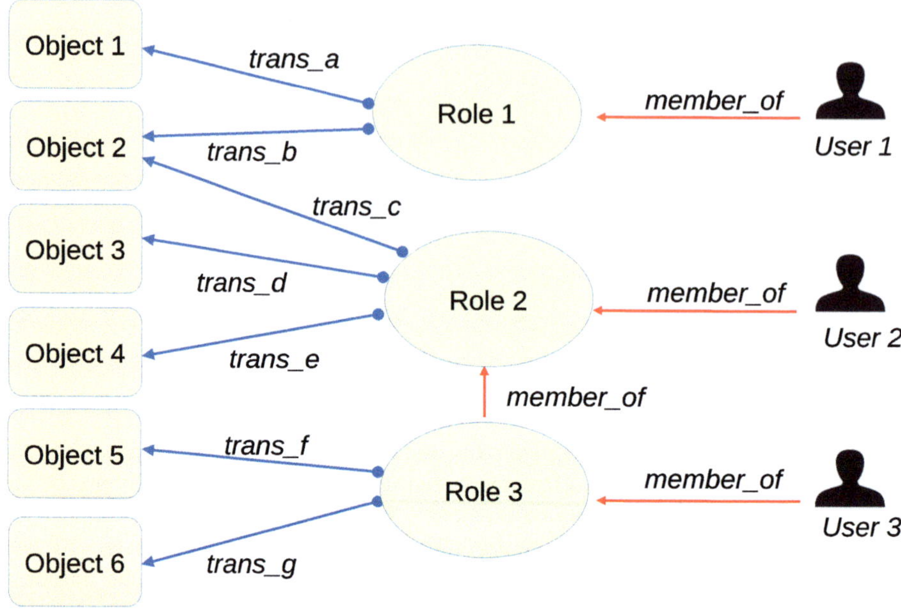

Fig. 12.1 Basic principle of Role-Based Access control with notation as suggested by Ferraiolo and Kuhn [1]

e.g., Role 3 in Fig. 12.1 being member of Role 2). In a ground segment context the definition of the following basic roles could be meaningful:

- The *operator* role being able to launch operations specific application software processes on a GCS element. A more specific definition could distinguish between operators doing flight dynamics, mission planning, mission control (this would include the release of TCs to the satellite), or ground segment supervision (monitoring and control).
- The *maintainers* role having the necessary access rights and permissions to configure, launch, and monitor typical maintenance tasks like the execution of regular antivirus scans, the update of the antivirus definition files, or the launch of backup processes.
- The *administrators* role with root level access at operating system level having the ability to install and remove software or change critical configuration files like for example network settings. The user access and profile management would also fall under this role.

In order to avoid the need to manage the user accounts on each machine in a ground segment separately, it is recommended to implement a centralised user management system. For this, well established protocols are available, with the most prominent one being the *Lightweight Directory Access Protocol* (LDAP) that allows the exchange of user profile information via an internet protocol (IP) based network.

This requires the setup of an LDAP server which supports the authentication requests from all the LDAP clients that need to access it. From a security point of view, it is recommended to establish the client to server connection via an encrypted connection, e.g., using the transport layer security (TLS) authentication protocol.

A further advantage of a centralised user management system is the possibility to store highly sensitive information like user passwords in one place only, where increased security measures can be applied. An important measure is the implementation of cryptographic hash functions like secure hash algorithms which avoid the need to store passwords in cleartext.[1] In this case the user password is immediately converted into its hashed form and compared to the encrypted one stored in the password file. Thanks to the nature of the hashing algorithm, it is not possible to reverse engineer the password from the hashed form, in case the password file gets stolen. However, strong hashing algorithms need to be applied (e.g., SHA-256, SHA-512, Blowfish, etc.) in order to reduce the risk of cracking software to be successful. Furthermore, a centralised user management system eases the definition and enforcement of a segment wide password policy which defines rules for a minimum password length, complexity, and validity.

12.2   File Transfer

A reliable and efficient exchange of files between the various elements is an essential requirement to ensure the proper functioning of the ground segment. The deployment of a capable file transfer mechanism is therefore an important aspect in the definition of the overall ground segment architecture. A very simple solution is to rely on the basic file transfer protocols provided by the operating system, which could be the *ftp* protocol or the more secure *sftp* variant that also provides data encryption. More advanced data transfer tools can however provide additional valuable features, like the ability to schedule automatic file deliveries including the initiation of action procedures (upon file reception) that can further process the received data.[2] Another useful functionality is the implementation of a monitoring functionality that provides feedback on the network status and the successful file transfer by the intended recipient. File transfer services can also ease the definition of routing tables which is useful if firewalls are deployed between elements.

Having mentioned some of the advantages that file transfer services can bring, one needs to also consider the potential risk of introducing additional complexity into a very sensitive area of the overall system. A failure of a file transfer system component could immediately affect a significant number of interfaces at the same

---

[1] Hashing is a mathematical method to produce a fixed length encoded string for any given string.

[2] An examples for an action procedures could be the initiation of data validation process (e.g., xml parsing, integrity check, antivirus check, etc.) and, depending on the outcome, a subsequent moving of it into a different directory.

time and lead to a major malfunctioning. In a worst case scenario, a complete outage of the entire ground segment simply due to the failure of the file transfer system could be possible. Especially multi node file transfer architectures may require the proper set up of a large number of configuration files which need to be properly tested during a dedicated segment qualification campaign.

12.3 Configuration Management System

Every ground segment needs to perform regular software updates in order to incorporate new functionalities, correct anomalies, or deploy the most recent cyber security patches. It is therefore recommended to integrate a dedicated Configuration Management System (CMS) into the ground segment architecture which provides the following main functionalities:

- A repository that hosts verified software releases, which are correctly configured and ready for deployment.
- A tool that is able to transfer software to a target machine and install it.
- A means to monitor the integrity of a software file or an entire directory structure and to identify and document any modifications with respect to an identified baseline.
- A gateway providing the operator a means to import and export data to the ground segment ensuring that adequate security measures are enforced. Such measures should as a minimum comprise an automatic antivirus screening for incoming files and the proper logging of any file transfers (e.g., which user has exported/imported what data at which time).

An example for a simplified CMS architecture is shown in Fig. 12.2 where the import/export gateway is also used to transfer an element software release from the

Fig. 12.2 Simplified example architecture of a Configuration Management System (CMS)

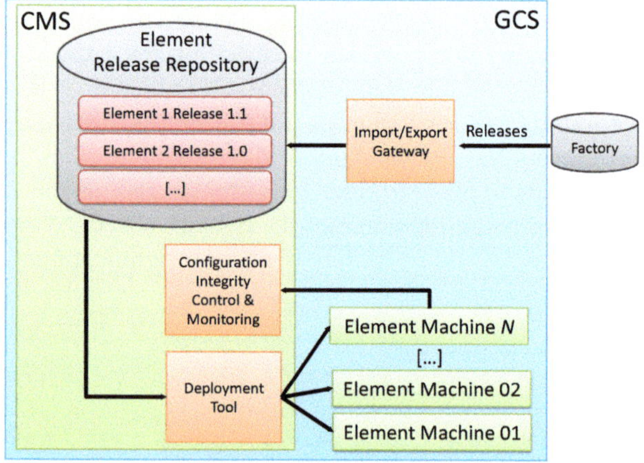

factory (where it has been developed and qualified) into the CMS release repository. The deployment tool has to interface that repository in order to copy and install a release candidate to a target machine. The configuration and integrity monitoring tool can be understood as a scanning device which performs a regular screening of the overall system configuration and flags any changes. This not only provides an efficient means to provide the overall segment (as-built) configuration but also helps to identify any unauthorised modifications.

In case of a critical system break down or an infection with a malicious virus (which cannot be isolated and removed), the CMS can serve as an important tool to perform a recovery of parts or even the entire ground segment software within a reasonable time frame.

12.4   Antivirus Protection

An antivirus (AV) protection system is an important function that needs to be present in every ground segment design. Every AV application typically consist of two main components, the scan engine and a database of known viruses. The latter one is also referred to as the *antivirus definition file* and requires a regular update to ensure the ability to detect the latest known viruses.

The deployment of the AV protection software can be realised in a decentralised or a centralised architecture. In the decentralised case, the engine and AV definition file are deployed at each single machine or instance where a scan needs to be performed, whereas in the centralised architecture they are only installed at one central server. In the latter case, any scanning, reporting, and virus isolation activities is then also only initiated from the central machine. The obvious advantage of the centralised version is the need to only update the AV definition file at one single location which avoids additional network traffic and allows an easier monitoring and control of the upgrade policy. A decentralised architecture might however be unavoidable for a remote site like the TT&C stations where the network connection bandwidth to a centralised server might be limited.

12.5   Data Protection

The objective of a data protection or backup system is to allow a fast and complete system recovery in case of data loss. As data loss can in principle occur at any time and potentially cause a major system degradation, a data protection functionality should be part of every ground segment design. The following types of data need to be distinguished and taken into account when configuring the backup perimeter of a data protection system:

- Configuration data which comprise all items that are subject to a regular change resulting from the daily use of the system and are distributed over the various software applications deployed in the entire ground segment. Furthermore, all databases (e.g., the full set of configuration items maintained by Operations Preparation Facility as described in Chap. 9) or configuration files of the various operating systems fall into this category. The complete list of datat items and their location must be clearly identified to ensure that they can be properly considered in the backup plan of the data protection system.
- System images can either represent an entire physical hard disk (e.g., ISO image files) or Virtual Machines (VMs) which have a specific format referred to as OVA/OVF files. It is recommended to generate updated system images at all relevant deployment milestones, like the upgrade of an application software version or an operating system.
- Operational data can be generated inside the ground segment (e.g., mission planning or flight dynamic products) or originate from an external source. Examples for external data sources are the satellite telemetry, the SRDB, or ground segment configuration items received from external entities.

The detailed design and implementation of the data protection system is very project specific but will in most cases rely on a backup server that hosts the application software and the data archive which needs to be adequately dimensioned for the project specific needs (e.g., expected TM data volume, number of software images, estimated size of configuration files, etc.). Backup agents are small software applications deployed on the target machines and help to locally coordinate the data collection, compression, and transfer to the main application software on the server. The data protection system should allow to schedule automatic backups, so they can be performed at times when less operational activities are ongoing. This will reduce the risk of performance issues when backup activities put additional load on the network traffic, memory, and CPU resources. Manually initiated backups should still be possible so they can be initiated right after an upgrade or deployment activity.

The definition of backup schedules and the scope of backups is often referred to as the segment *backup policy* and needs to be carefully drafted to ensure it correctly captures any relevant configuration change that is applied to the system. At the same time, the policy should minimise the data footprint in order to avoid the generation of unnecessary large data volumes which need to be stored and maintained. A possible strategy to achieve this, is to schedule small size differential backups on a more frequent basis[3] targeting only a subset of files, and full disk images whenever a new version of an application software or operating system is deployed (see Fig. 12.3). With this strategy in place, a full system recovery should always be possible, starting from the latest available disk image and deploying differential backups taken at a later time on top.

---

[3] Differential backups should be scheduled as a minimum each time a configuration change has been applied.

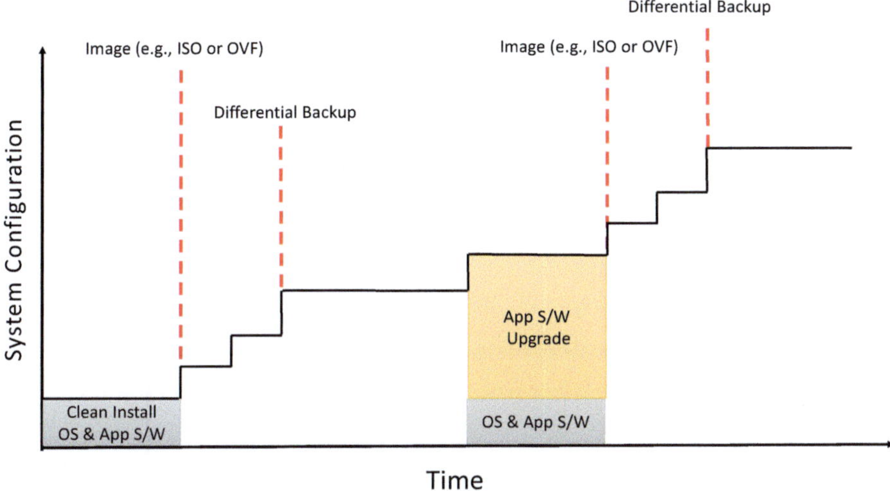

Fig. 12.3 System configuration as function of time: a possible backup policy could consider the generation of images after major upgrades and differential backups to capture smaller configuration changes

Finally, it is worth to stress that any backup of data should also provide a means to check the integrity of archived data to ensure that no errors have been introduced during the backup process (e.g., read, transfer, or write error). An example for a very simple integrity check is the generation and comparison of checksums before and after a data item has been moved to the archive.

12.6   Centralised Domain Names

Even a simple ground segment architecture needs to deploy a considerable amount of servers and workstations which are connected via a network using the internet protocol. This requires the configuration of internet protocol (IP) addresses and domain names which map to the hostnames of the connected devices in a local network. The mapping of IP addresses to the corresponding hostnames can either be stored in a local configuration file that is part of the element OS[4] or be managed by a centralised Domain Name System (DNS) server that provides this information to the DNS clients. Due to the relevance and sensitive nature of this type of information, a DNS server should be protected both from a security point of view (e.g., hardened operating system) and from a failure point of view through the provision of redundancy.

---

[4] E.g., the /etc/hosts file in Unix/Linux based systems.

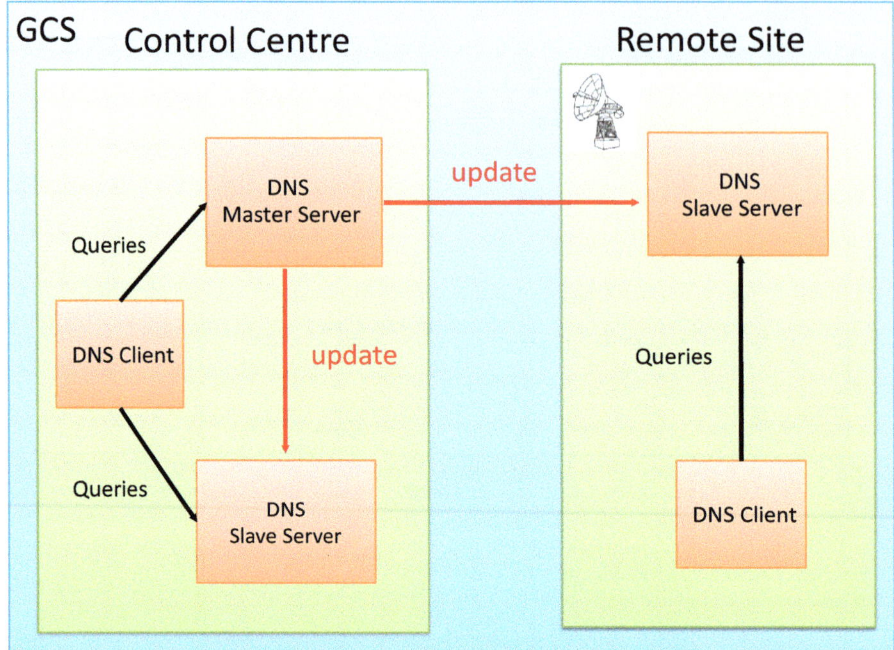

Fig. 12.4 Schematic outline of a centralised and redundant DNS server configuration

An example of a redundant and centralised DNS sever setup is shown in Fig. 12.4 which consists of a DNS master and a slave server. The slave is synchronised with the master on a regular basis, and is ready to take over in case the master fails. Due to the geographical distance of remote sites (e.g., a TT&C station on a different continent), the limiting bandwidth and reliability of the network connection favours the deployment of a local DNS slave server which is also shown in Fig. 12.4. The local DNS server is used for the daily operation of the remote DNS clients, and the wide area network connection is only needed to synchronise the DNS configuration between the remote slave and the master which is located in the main control centre. As the data volume for a DNS configuration is very small, the required network bandwidth is very low.

12.7   Time Source and Distribution

All servers and work stations of a ground segment need to be synchronised to a common time system. The time source can be either located in the ground segment or derived from an external source. Precise timing sources are usually part of ground segments that operate navigation satellite constellations (e.g., Galileo, GPS,

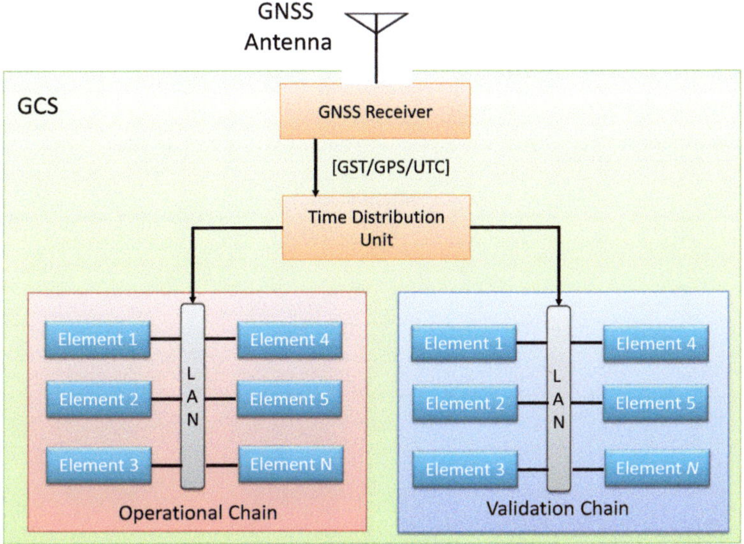

Fig. 12.5 Time synchronisation inside the GCS based on the reception of a GNSS based time source

GLONASS, etc.) where a highly complex and expensive atomic clock on ground is a mandatory asset to operate the navigation payload. Such a clock will usually not be available for other projects which therefore have to rely on an external source. With the availability of several operational navigation services today, the deployment of a GNSS antenna and a corresponding receiver is a very simple means to acquire a highly accurate and stable external time source as shown in Fig. 12.5. The same navigation signal also provides all the relevant information to derive UTC which is a very common time system used in space projects.

The time signal can be distributed inside the GCS via a time signal distribution unit that implements standardised time synchronisation protocols like the *Inter-Range Instrumentation Group* (IRIG) time codes or the *Network Timing Protocol* (NTP). The time distribution could also be done to remotely located TT&C stations, using the wide area network connection. Alternatively, a local GNSS antenna could be deployed at the remote site to avoid the dependency on the network.

Reference

1. Ferraiolo, D. F., & Kuhn, D. R. (1992). Role-based access controls. In *Proceedings of the 15th national computer security conference (NCSC), Baltimore, Oct 13–16, 1992*, pp. 554–563.

Chapter 13
The Physical Architecture

Whereas in the previous chapters the functional aspects of the ground segment where described, this one focuses on the physical implementation of the infrastructure. This comprises the actual hardware (workstations, servers, and racks) that need to be deployed but also the setup and layout of the server and control rooms hosting it. Furthermore, the virtualisation technology is introduced which is a very important technology that allows to significantly reduce the hardware footprint through a more optimised use of the available computing resources. Virtualisation also provides an efficient means to add and manage system redundancies and improve the overall segment robustness which is an important feature to ensure service continuity to the end users of a system. Finally, some considerations are introduced to help the planning and execution of large scale system upgrades or even replacements which is referred to as system *migrations*.

13.1 Client Server Architecture

The client server architecture is the preferred computing model deployed in ground segments as it provides many advantages in terms of data and resource sharing, scalability, maintenance, and secure access control. The basic concept is illustrated in Fig. 13.1 where the operator accesses the application software via a client workstation that is placed at a different location (e.g., a control room) and connected to the server via the local area network (LAN). As the application software and its data are physically deployed and run on the server machine, they have much higher needs for processing power, RAM, and data storage volume and are therefore usually mounted in racks and deployed in server rooms. A network attached storage (NAS) can be used to keep all operational data which can also implement some

B. Nejad, *Introduction to Satellite Ground Segment Systems Engineering*, Space
Technology Library 41, https://doi.org/10.1007/978-3-031-15900-8_13

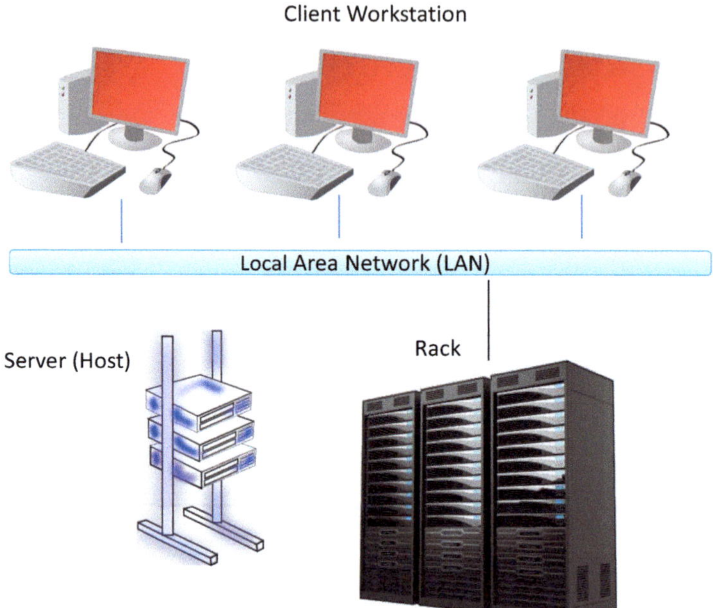

Fig. 13.1 Illustration of a client Server architecture

kind of data storage virtualisation technology to protect from data loss in case of disk failure.[1]

Compared to the server, the client workstations can have a much lighter design and therefore need less space and generate less heat and noise. These are therefore more suitable to be accommodated in office or control rooms, where humans work and communicate. Client workstations can be either dedicated to a specific server or element or be independent which means that any client can be used to operate any element. If the client workstations do not have any software or data installed locally, they are commonly referred to as a *thin* client architecture.

Servers can be grouped and mounted into racks which are set up in rows in dedicated server rooms as shown in Fig. 13.4. With multiple servers mounted in one rack operating on a 24/7 basis, a considerable amount of heat is generated that needs to be properly dissipated in order to avoid hardware damage. With several racks usually hosted in the same room the appropriate sizing of the air conditioning capacity is an important aspect for the server room design and will be discussed in more detail in Sect. 13.4.

---

[1] A frequently used storage virtualisation concept is the *Redundant Array of Inexpensive Disks* or RAID architecture which provides a variety of configurations (i.e., RAID 0 to 6) with each of them having a unique combination of disk space, data access speed, and failure redundancy [1].

13.2 Control Rooms

The flight control team (FCT) performs all the satellite operations from a dedicated control room in which the client workstations (consoles) are located. Quite commonly, a ground control segment architecture provides several control rooms with different size, layout, and hardware equipment. The largest one is usually referred to as the *main control room* (MCR) and is used during important mission phases like the LEOP, major orbit relocation activities, or contingency operations with intense ground to satellite interaction during which the full FCT team capacity needs to be involved.

Not all satellite operational activities require a real time ground-to-space interaction but focus on offline work like the generation of flight dynamics products or the planning of time lines. Such activities do not require to be to be done in the same rooms where the FCT is located, but can be better transferred into specific areas dedicated for planning, offline analysis, and flight dynamics.

Once a satellite project goes into routine phase, the necessary ground interaction will usually decrease and so the team of operators that need to be present. During this phase, satellite operations can be transferred into a smaller sized room which is referred to as *routine or special operations room* (SOR).

13.3 Chain Separation Concept

In order to provide operators an isolated environment for the validation of flight control procedures and to perform all the training activities without impacting ongoing flight operations, a network or *chain separation concept* should be implemented. This concept foresees a duplication of all ground segment components on two separate network environments, referred to as operational (OPE) and validation (VAL) chains, which must be strictly isolated from each other as shown in Fig. 13.2.

The OPE chain is used for real satellite interaction and must therefore be connected to the TT&C station network. The VAL chain is only used for validation, simulation, or training campaigns and should therefore be isolated from the station network in order to avoid any unplanned satellite communication. To ensure a representative environment for the validation of flight procedures, any TM/TC exchange with the satellite is replaced by the satellite simulator which hosts and executes the OBSW as explained in Chap. 11.

To ensure a representative behaviour of the ground segment, the configuration of all elements and components (e.g., versions of installed application software, database contents, network configuration, etc.) between the OPE and VAL chains must be fully aligned. In the frame of a System Compatibility Test Campaign (SCTC) aiming to demonstrate the ground segment ability to communicate to the space segment (prior to launch), the VAL chain is connected to the satellite via the electronic ground support equipment (EGSE) located in the *Assembly Integration and Test* (AIT) site (see lower part of Fig. 13.2).

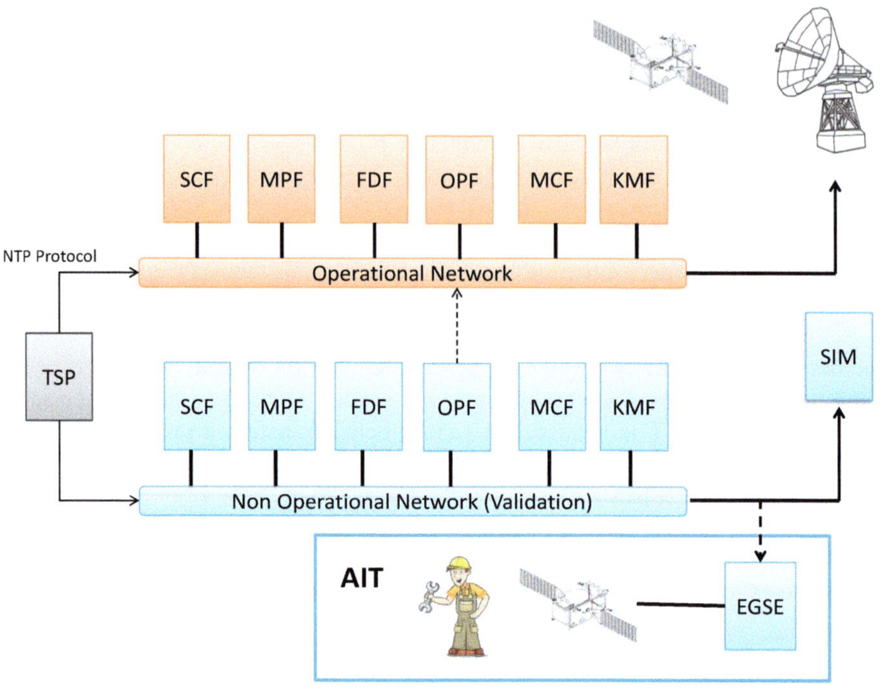

Fig. 13.2 Chain separation concept: the operational and validation chains are deployed on separated networks. The time source provider (TSP) is conceptualised here as a single entity connected to both networks, but alternative implementations (e.g., with separated components) are also possible

13.4   Server Room Layout

Server rooms need to host a large amount of very performant computer hardware mounted in server racks which are usually arranged side-by-side and in rows in order to optimise the use of available space. The close proximity of servers inside a rack and the racks themselves imply a considerable generation of heat that needs to be measured and controlled in order to avoid a hardware overheating that could reduce its lifetime or even lead to damage. The monitoring and control of the rack temperature can be done with the following means:

- To manage the heat dissipated by the servers, a ventilation system is mounted inside the rack that generates an air flow of cold air entering the rack and hot air leaving it (see ventilator symbol at the top of the rack in Fig. 13.3). The cold air can either enter from an opening at the bottom (e.g., from below the false floor) or from the front through perforated doors. The hot air leaves the rack from the rear side.

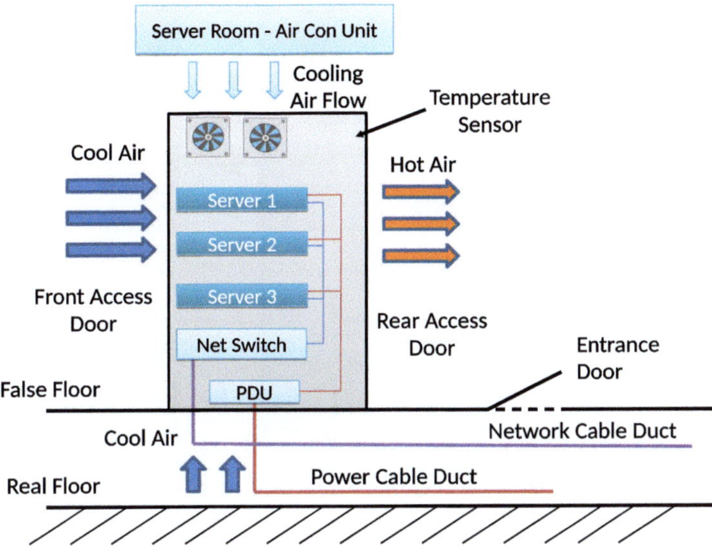

Fig. 13.3 Illustration of a rack with three servers mounted on a false floor. The power distribution unit (PDU), network switch, ventilators, and temperature sensor inside the rack are indicated. Also shown are the cabling ducts for the network and power source below the false floor and the server room air conditioning unit below the ceiling

- Each rack is equipped with a temperature sensor which sends an alarm to a monitoring system in case a maximum limit is exceeded. In such a case, an immediate investigation should be initiated and corrective actions taken in order to avoid potential equipment damage.
- The server room itself must also be temperature controlled with its own air flow and conditioning system deployed. This system needs to be properly sized according to the number of hosted racks and their consumed power and also consider room specific properties that can influence the temperature distribution inside.
- Taking into account the rack specific heat flow, i.e., cool air entering from the front, hot air leaving at the back, the arrangement of the adjacent rack rows should support the correct arrangement of cold and hot isles. The examples presented in Fig. 13.4 show an inefficient thermal layout in the upper panel that results in a pile up of hot air at the front side of the racks, and a more efficient one in the lower panel with the front sides always facing a cold and the rear ones a hot isle.

13.5   Rack Layout, False Floor, and Cabling

Each rack has to be connected to the power supply of the building which usually provides a no-break and a short-break power source. The no-break source is protected via an uninterruptible power supply (UPS) unit which can guarantee a

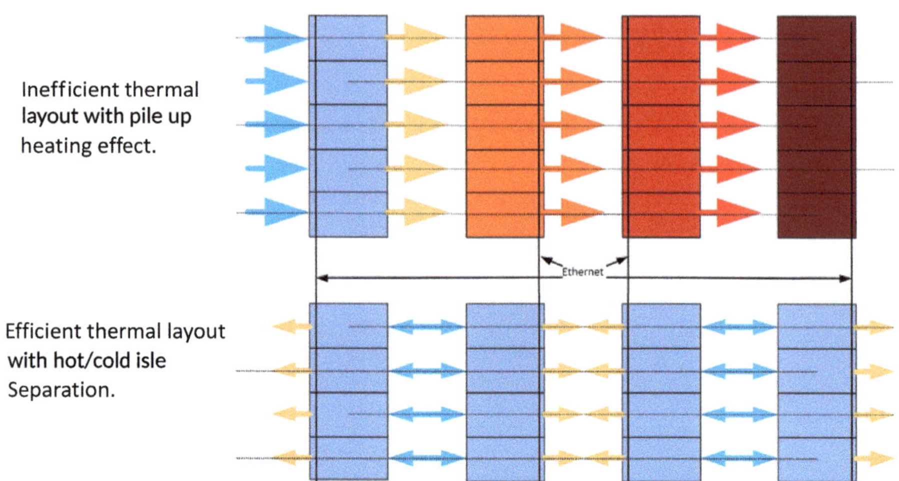

Fig. 13.4 Examples for rack arrangement inside a server room. The upper example shows a suboptimal configuration that implies a pile up heating from one row to the next, whereas the lower arrangement supports the availability of cool air from the front for all rows (hot/cool isle separation)

continuous supply in case of a power cut or outage. As the UPS is a battery based unit, it can only replace the main source for a relatively short time period (usually less than one hour). To protect against longer outages (e.g., several hours up to days), a Diesel generator with an adequate tank size is more suitable. Such a generator should either be an integral part of the ground segment building infrastructure or at least be located somewhere in close vicinity to allow a connection to the ground control segment.

Sever rooms that host several rack positions require a large amount of cables to be routed throughout the room in order to connect each position to the desired end point. This is most efficiently achieved through the deployment of a *false floor*[2] as shown in Fig. 13.1. When planning the cable routing, it is important to separate network and power cables to avoid interference. Usually a large amount of network cables (also referred to as patch cables) have to be deployed next to each other and over long distances which can cause crosstalk that reduces the signal quality. It is therefore recommended to choose cable types that are designed to minimise such crosstalk. This is achieved through a twisting of wire pairs with different current directions inside the cable which will cancel the overall fields generated by these wires. Patch cables can combine several twisted pairs in one cable ensuring that no twists ever align. Additional foil shielding around each twisted pair will improve the performance over long distances where cables usually pass through areas of high electrical noise. The quality of shielding of a cable can be derived from its specified

---

[2] A false floor is sometimes also referred to as raised or access floor.

categorisation (e.g., Cat 5, Cat-6, Cat-6a, etc.) which provides an indication on the maximum supported transmission speed and bandwidth. The choice of the right cable category is therefore an important consideration for the deployment of large scale network and computing infrastructure.

The distribution of hardware inside one rack requires careful planning in order to avoid unnecessary difficulties for its maintenance or upgrade at a later stage. For maintenance activities, it is very important that the racks are arranged in a way that allows a person to access them from the front and rear side.

13.6 Migration Strategies

Throughout the lifetime of a project, a larger scale upgrade of the entire ground segment might be required. Possible reasons for this could be the obsolescence of server and workstation hardware requiring a replacement, or the implementation of design changes due to the addition of new functionality. Larger scale upgrades always imply an outage of a larger portion or even the entire ground segment. The purpose of a *migration strategy* is to define and describe the detailed sequence of such an intervention, in order to minimise the infrastructure down time and minimise the impact on the service. This strategy must of course be tailored to the scope of the upgrade and the very specific design of the ground segment and requires detailed knowledge of the hardware and software architecture.

Two high level migration concepts are outlined below which can serve as a starting point for a more detailed and tailored strategy. Whichever concept is considered to be more suitable, it is always important to have a clear and detailed description for both the roll-out and a roll-back sequence in place. Especially the latter one is an important reference in case major problems occur during the migration. Having a clear plan on how to deal with potential contingencies including a detailed description of how to overcome them, will provide more confidence to managers and approval boards, who have to authorise such a migration in first place.

13.6.1 Prime-Backup Migration

The *prime-backup* migration strategy requires the existence of two separate ground control segments which are designed and configured to operate in a fully redundant mode (see Fig. 14.4). This means that each of the two segments must be capable to operate the space segment in isolation, without the need to communicate with the other centre. This implies that both segments are connected to the remote TT&C station network and any other external entity required to operate the satellites. For such a setup, a possible migration strategy is described in the following four stages which are also graphically shown in Fig. 13.5:

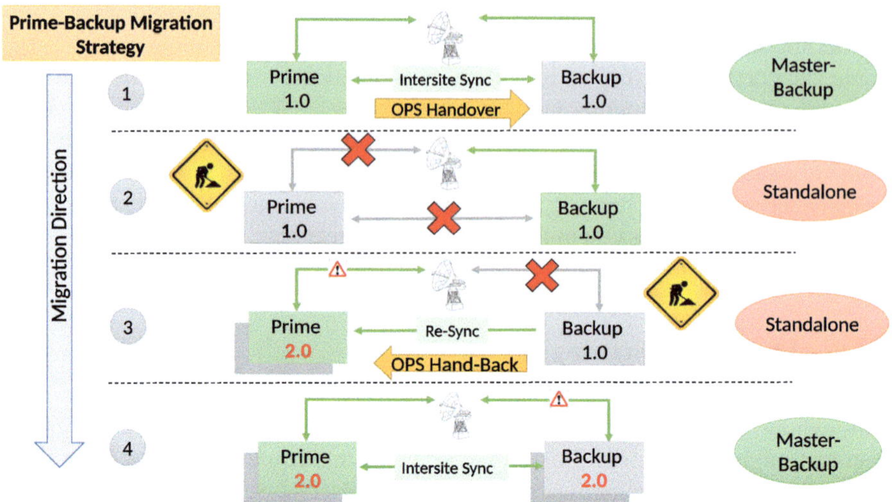

Fig. 13.5 Outline of the *prime-backup* migration strategy steps

1. In the nominal configuration the prime and backup sites operate in the so called prime-backup configuration, which means the prime is in charge of satellite operations and performs a regular inter-site synchronisation of all operationally relevant data to the backup centre. Once the readiness for migration has been confirmed (usually by a checkpoint meeting), the *planned* handover[3] of operations from the prime to the backup site can be initiated. This is done following a detailed handover procedure and requires a detailed and advanced planning as it might require a transfer of operational staff to the backup site.

2. Once it is confirmed that the space segment can be successfully operated from the backup site (without the need of the prime), both sites can be configured into *standalone* mode which stops the inter-site synchronisation. Now the prime segment can be isolated from the external network connection which implies an intentional loss of contact to all remote and external sites from this centre. After successful isolation of the prime, all required upgrade activities can now be performed without any operational restrictions which is indicated by the transition from GCS v1.0 to v2.0 in Fig. 13.5.

3. Once the prime upgrade has been finalised, the internal interface connections can be reestablished in order to resynchronise any new data to the prime site. As a next step, the reconnection of the external interfaces can be performed. This might however not be trivial, in case the network equipment has been upgraded at the prime site and is now connected to old network equipment at the remote sites. In case backwards compatibility is not supported, a reconfiguration

---

[3] There is also the concept of an *unplanned* handover which needs to be executed in an emergency case and follows a different (contingency) procedure that allows a much faster transition.

of network devices at the remote site might be required, which is indicated by the small warning sign on the interface line in Fig. 13.5. This could require the reconfiguration of IP addresses, upgrade of firmware in a network switch, exchange of VPN smartcards, or even the replacement of the network equipment. It is highly recommended to perform a thorough testing of the upgraded segment at this stage. Such a validation could also involve a test contact to the space segment in the form of a listening mode that only receives telemetry but does not transmit telecommands.[4] Once full operational functionality of the upgraded prime segment has been confirmed, the hand-back of operations to the prime can be initiated. Once the prime has regained full operational responsibility, the isolation and consecutive upgrade of the backup site can be initiated.

4. Once the upgrade of the backup site has been finalised, the inter-site connection to the prime segment can be re-established in order to ensure full data synchronisations between the two segments. This is followed by the connection of the backup site to the external entities. At this point, full prime-backup redundancy has been reestablished with both segments upgraded. The migration activity can now be considered as completed.

13.6.2 Bypass Migration

The *Bypass* migration strategy is the preferred option if no redundant backup segment is available that can take over satellite operations and allow an isolation of the infrastructure to be upgraded. This is the main difference compared to the previously described prime-backup migration strategy and it is obvious that the planning of such an intervention is considerably more involved and also bears a much higher risk. During the migration, adequate measures must be taken to ensure a continuous ability of the ground segment to command and control the space segment.[5] To some extent, this scenario resembles an open-heart surgery in which the patient is put on a cardiopulmonary bypass that replaces the function of the heart during the time it undergoes surgery (refer to e.g., [2]). In a similar manner, the ground segment migration methodology described here relies on the implementation of a temporary *bypass infrastructure* which is able to take over specific ground segment functions to allow the underlying infrastructure to be taken out of service. Similar to the prime-backup strategy, the bypass method can be phased into four major steps which are graphically shown in Fig. 13.6 and described below:

---

[4] The transmission of a test command with no impact (e.g., a ping TC) could still be considered as it does not interfere with ongoing operational activities.

[5] Even if a satellite can in theory survive a certain amount of time in space without ground segment interaction, this should be avoided as it usually implies a degradation or even interruption of the service it provides.

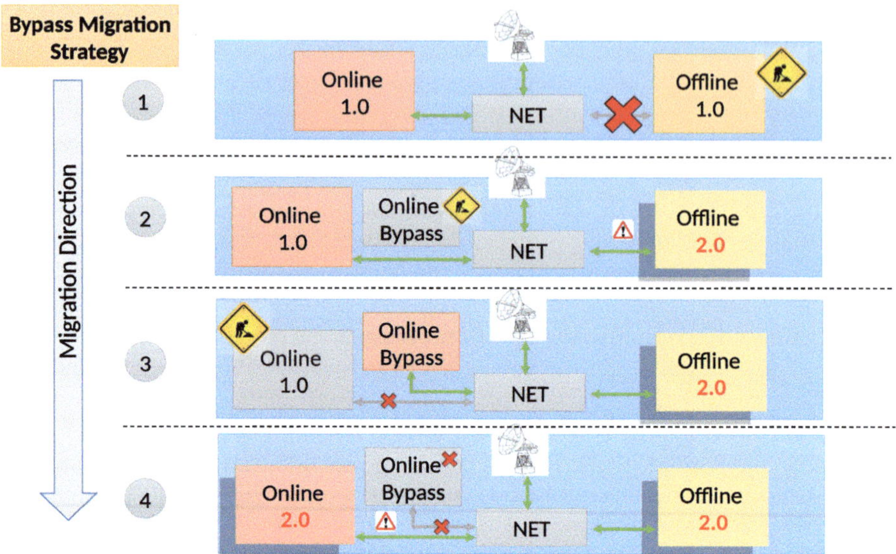

Fig. 13.6 Outline of the *Bypass* migration strategy steps. NET = segment network infrastructure

1. In the first step the ground segment is grouped into elements providing *offline* and *online* functionality. The offline elements are not directly required during the execution of a satellite contact as they mainly contribute to the preparation of operational products. Typical elements belonging to this category are the Flight Dynamics Facility (see Chap. 6), the Mission Planning Facility (see Chap. 8), the Operations Preparation Facility (see Chap. 9), or the Satellite Simulator (see Chap. 11). Therefore, the timing of their upgrade can be readily coordinated with their required product delivery time. In an extreme case, they could even be upgraded during a running contact without any major impact, which should however not be considered as good engineering practice. In the example given here the offline element upgrade is done at the very start of the overall migration schedule, but it could in principle also be done at the very end. It is however recommended to separate the upgrade of the offline elements from the online ones, in order to reduce the risk of multiple failures occurring at the same time and simplify trouble shooting in case of problems.

2. In the second stage, the implementation of the bypass infrastructure is performed which comprises the deployment of the necessary hardware, the installation of the required application software, the connection to the segment network, and the proper configuration of operational data (e.g., installation of the satellite specific TM/TC database, up-to-date Pluto procedures, etc.). At the end of this phase, the readiness of the bypass infrastructure to take over the critical online functionality (e.g., the TM/TC exchange with the satellite) needs to be proven. This could be done via the execution of a simple a test contact during which the

bypass infrastructure is run in parallel, performing telemetry listening and only non critical commanding (e.g., ping command).

3. The third phase is a critical one, as the online elements are taken out of service and disconnected from the network. Once the bypass infrastructure has taken over their functionality, the disconnected elements can be upgraded or replaced. A thorough qualification campaign needs to demonstrate the full and proper functionality of the new items.

4. In the fourth phase the upgraded infrastructure can be put back into service and re-connected to the network in order to replace the bypass infrastructure. The latter one can then be disconnected, but should be kept in place for a certain amount of time, so it can be quickly reactivated if needed.

References

1. Patterson, D. A., Gibson, G., & Katz, R. (1988). A case for redundant arrays of inexpensive disks (RAID). In *Proceedings of the ACM ACM SIGMOD international conference on management of data*, pp. 10–116. ISBN: 0897912683, https://doi.org/10.1145/50202.50214.
2. Cohn, L. H. (2017). *Cardiac surgery in the adult* (5th ed.). McGraw-Hill Education-Europe. ISBN 13 9780071844871.

Chapter 14
Virtualisation

Hardware virtualisation technologies have improved significantly over the past decades, and have reached a level of maturity today, making them an omnipresent building block in all modern cloud centre architectures. Virtualisation is also a core feature in modern satellite ground segment engineering and a key enabler to modernise existing infrastructure and deal with obsolete hardware and heritage operating systems. A virtualised architecture provides many benefits that improve the overall system performance and at the same time allows a significant reduction in hardware footprint through a more efficient use of the available resources.

14.1 Hyper-Converged Infrastructure

The basic architecture of a classical (i.e., non virtualised) computer system is shown Fig. 14.1. At the very bottom sits the hardware layer which can be a rack mounted server or a smaller device like a workstation or a simple desktop computer. The next higher layer is the Basic Input Output System (BIOS)[1] which is the first piece of software (or firmware) loaded into the computer memory immediately after the system is powered on and booting. The BIOS performs hardware tests (power-on self tests) and initiates a boot loader from a mass storage device which then loads the operating system (OS). Once the OS is up and running, one or several user applications can be launched and operated in parallel (indicated by "App-1,-2,-3" at the top of Fig. 14.1).

A basic example of a virtualised architecture is shown in Fig. 14.2. The hardware layer at the very bottom is again followed by the BIOS or UEFI firmware which is loaded after the system is powered on. The next higher layer in a virtualised system

---

[1] The BIOS has been supplemented by the Unified Extensible Firmware Interface or UEFI in most new machines.

© The Author(s), under exclusive license to Springer Nature Switzerland AG 2023   199
B. Nejad, *Introduction to Satellite Ground Segment Systems Engineering*, Space Technology Library 41, https://doi.org/10.1007/978-3-031-15900-8_14

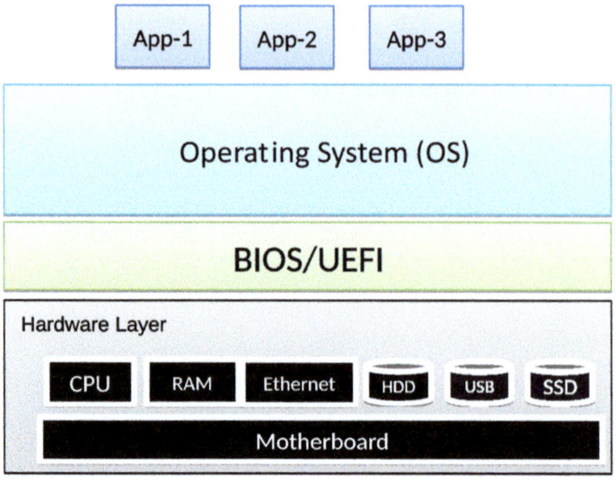

Fig. 14.1 High level architecture of a legacy (non-virtualsied) computer system

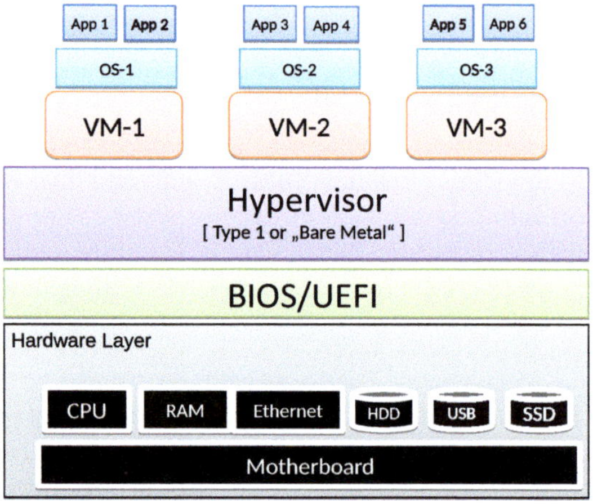

Fig. 14.2 High level architecture of a virtualised computer system

is however not the OS, but a hardware abstraction layer referred to as type 1 or bare metal *hypervisor*.[2] This hypervisor is able to create and run one or several *virtual machines* (VMs) which can be understood as software defined computer systems based on a user specified hardware profile, e.g., a defined number of CPU cores, memory size, number of network adaptors, disk controllers, or ports. Each VM provides the platform for the installation of an operating system (e.g., Linux, Windows, etc.), which serve as a platform to install the end users applications.

Hypervisors provide virtualisation functions in three IT domains, computing (CPU and memory), networking, and storage which are briefly outlined below. An infrastructure that implements all three components is referred to as a *hyper-*

---

[2] The difference of a type 2 hypervisor is that it is launched from an already installed operating system, similar to an application.

converged infrastructure or HCI and is state of the art in modern software-defined data centres.

14.1.1 CPU and Memory Virtualisation

Virtualisation allows to configure and launch a large number of VMs on the same physical server, with each having a different hardware profile and even running a different operating system. If VMs are operated in parallel, they have to share the hardware resources of their common host server and it is the task of the hypervisor to properly manage the CPU and memory resources among the VMs. There is however an important limitation that the maximum number of (virtual) processor cores (vCPUs) of a VM must always be lower than the physical number of CPU cores of the host machine.

14.1.2 Network Virtualisation

Virtualisation technology also provides efficient means to connect VMs in a virtual network using (software based) virtual network interface controllers (VNIC) which emulate the behaviour of their physical counterparts (see Fig. 14.3). The definition of a virtual switch not only connects a set of VMs among each other, but also allows to connect these to an external physical Ethernet adaptor of the host server. Modern versions of hypervisors also support advanced network switch features like the definition of virtual LANs (as per IEEE 802.1Q network standard), traffic shaping (e.g. limiting the consumption of network bandwidth for a single VM), NIC teaming with load balancing, network adaptor failure policies, and various security features.

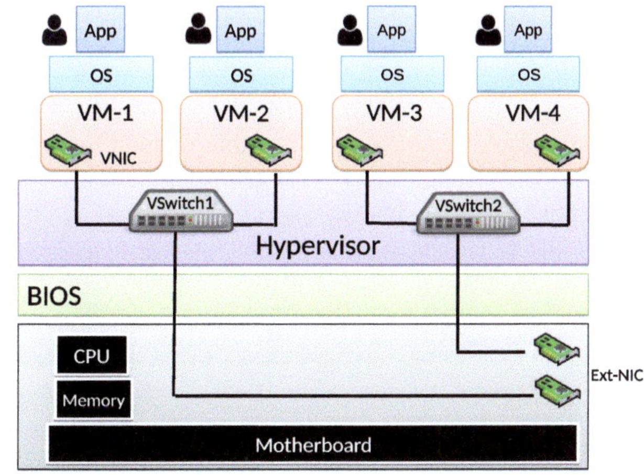

Fig. 14.3 Example of simple virtual network configuration. VM-x = Virtual Machines, VNIC = Virtual Network Interface Controller, VSwitchx = Virtual Switches, Ext-NIC = External (physical) network interface controller

14.1.3   Storage Virtualisation

A virtual SAN or vSAN is generated by grouping a set of physical storage devices like hard disks, solid state disks, or tape drives that are physically attached to their respective host machines to a virtualised shared storage area. This vSAN storage can be accessed by each of the VMs that is deployed on a hosting server which is logically connected to other servers in a so called *structure* arrangement. Such a software defined storage solution avoids the need to deploy complex network attached storage arrays with RAID controllers. Due to the reduced cabling requirements, the overall design can be simplified and the hardware footprint further reduced.

14.1.4   Hardware-Software Decoupling

An operating system (OS) can only make proper use of a hardware component if the hardware specific driver software for that operating system is available. Such drivers are usually provided and maintained by the hardware vendor. If an obsolete hardware item needs to be replaced by a newer model, the new driver software coming with it might not be supported by the old OS anymore.[3] In such a case, the obsolescence recovery will not be possible unless the OS is upgraded to a newer version. Doing so, some application software might however not work properly anymore, as it could depend on a set of libraries that only run on that old OS version. After the application software with all its dependencies (e.g., other COTS software products) has been ported, it needs to be requalified to ensure correct output and performance. This is even more relevant for a software running on a critical system that is subject to a specific certification or accreditation process, prior to its formal use.

The use of a virtualised architecture can help to avoid such an involved process of software porting, as the hypervisor acts as an intermediate layer between the OS and the hardware and can provide the necessary backwards compatibility to an older OS. To give a simple example, a hypervisor running on a brand new server hardware can be configured to run a VM with an old and obsolete OS (e.g., Suse Linux Enterprise 9). This would not have been possible without the legacy drivers that are provided by the virtualisation layer. This setup now allows to operate a legacy application software on exactly that (legacy) OS it has been qualified for but on new hardware. This concept is referred to as the decoupling of software from hardware and avoids the need to port application software to a newer OS version in order to benefit from new and more performant hardware.

---

[3] Try to install a brand new printer on a very old computer running Windows XP !

14.1.5   VM Management

The deployment of a virtualised ground control segment architecture requires the creation and management of a large number of VMs. The most common virtualisation hypervisors on the market today offer advanced centralised management tools from which all the VMs and their network connections can be created, configured, clustered, monitored, and even removed.[4] Such management tools can therefore provide a subset of functionalities of a Monitoring and Control Facility (MCF) as described in Chap. 10.

14.1.6   Redundancy Concepts

The consideration of hardware redundancy in a ground segment architecture allows to overcome component or system failures, and ensures the continuation of service provision being an important requirement for every critical infrastructure. There are two main redundancy concepts in use which are referred to as cold and hot redundancy. A cold redundant device is nominally powered off and only activated in case the prime unit fails. The draw back of such a configuration is the time or latency for the backup unit to be able to fully replace the failed prime. In a hot redundant configuration, the backup unit is always up and runs in parallel to the prime unit without performing any of its function. In case the prime fails, the redundant unit can take over in more a seamless and faster way. The drawback is a higher energy consumption and heat production of the entire system and an earlier wear out of the redundant device due to its continuous activation. Both cold and hot redundant configurations have their justification, and it is the responsibility of the system designer to find the right trade off between system robustness and cost.[5]

A hot redundant sever concept with two servers named "A" and "B" is schematically shown in Fig. 14.4. As a server failure can occur at any time without significant signs or warnings, it is necessary to deploy a dedicated network based application referred to as *heartbeat* that continuously monitors the presence and functionality of each server and provides this information to its counterpart. In case the prime fails, the backup server can then launch all the necessary processes and take over the prime's role. To avoid the transfer of large data volumes from A to B, an external storage array (or a virtualised storage) is used, which is accessible to both servers.

In ground segment engineering, redundancy concepts are not only defined at component or server level but also at the level of the entire ground control segment. Such a segment level redundancy can be realised through the deployment of a

---

[4] To give an example, the widely used bare metal hypervisor ESXi® of the Vmware Incorporation provides vCenter® as their management centre solution [1].

[5] System robustness should always be a prime objective aiming to avoid any single point failure by design.

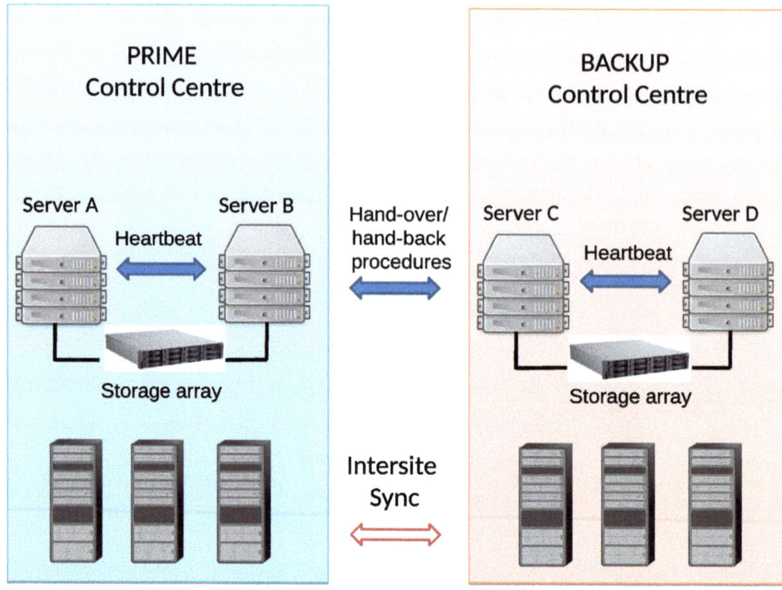

Fig. 14.4 Redundancy concepts at server level (Server-A and -B) and center level (GCS Prime and GCS Backup)

secondary (blue print) ground segment at a geographically different location. This allows to safeguard against a centre wide failure due to a catastrophic event (e.g., extreme weather conditions, earthquake, cyber attacks, etc.). As the deployment and maintenance of such a backup segment is very cost intensive, it might only be viable for larger scale projects that provide system critical services (e.g., GNSS or defence related services). In order to allow a seamless transition from one ground segment to its counterpart in a reasonably short time span, the design needs to foresee mechanisms of regular data synchronisation. It is furthermore required that both centres are connected to the remote TT&C station network in order to take over the satellite commanding and control responsibility.

Virtualisation also supports and simplifies the implementation of redundancy. The configuration of several servers to a server *cluster* allows the hypervisor to take over the monitoring role of the heartbeat application. If a host server in a cluster fails, the hypervisor can automatically replace the affected VMs by launching corresponding ones on an operational server. Such a configuration is referred to as a virtualised *High-Availability* (HA) cluster. Even if a HA cluster aims to minimise the downtime of the affected VMs, some loss of data or transactions cannot be avoided.

If no downtime and data loss can be accepted at all, the *Fault-Tolerant* (FT) configuration should be the preferred choice. In this case, each VM on one server is replicated by a second VM on a redundant server and is continuously synchronised through the mirroring of virtual disks. It is important to note that these two VMs represent a single VM state with the same network identity (e.g., IP address), despite the fact that they are physically separated. This allows a seamless transition with no down time in case of failure of the physical server where the primary VM is running.

14.2 Containers

In the traditional virtualisation concept as described in Sect. 14.1, a hypervisor virtualises physical hardware by generating a VM, which then provides the platform for the installation of the guest OS. In other words, the VM is a software defined computer that makes its guest OS believe to run on real hardware. This allows to run several VMs with different guest OS (e.g., Linux, Windows, and Mac OS) on the same host in parallel, which is shown on the left side of Fig. 14.5. Despite the various benefits pointed out in Sect. 14.1, the VM based concept has also weaknesses which are briefly summarised below.

- Each VM must include a full copy of the OS it is running on with all the required binaries and libraries. This can be resource heavy and take up significant space both at hard disk and memory level, especially if a large number of VM must be deployed.
- In order to run an application that is used for a specific service, the corresponding VM it is installed on must first undergo a full OS boot sequence and this takes time. While booting a VM can take several seconds, the start of a container goes as fast as a Linux process (order of milliseconds) [2].
- Applications like database servers work with block storage that require direct access to hardware resources (e.g., disks, network devices). Each access request must always be routed through the guest OS, the VM abstraction layer, and the hypervisor, which can slow down the application performance [3].

Container virtualisation offers an attractive alternative that overcomes these weaknesses and also provides a separation of environments for applications running on the same host system. Instead of virtualising the underlying hardware, containers virtualise the operating system (typically Linux or Windows). Each individual container hosts the application and the required OS dependencies and libraries which can access the host OS kernel directly. Containers are therefore small, fast, and (unlike a VM) do not need to include a complete guest OS in every instance.

Thanks to the presence of application and OS dependencies inside a container (indicated as "Virtualised OS Lib + Bin" in Fig. 14.5), the application can be easily ported between different hardware environments with different host operating system distributions. This makes containers very attractive for software development for which the final target environment is not clearly defined at the time of development, which is of special interest in a cloud computing context [4]. As the (application) code inside a container will have the same behaviour no matter what computer it is finally operated from, containers are also very efficient from a software testing point of view.

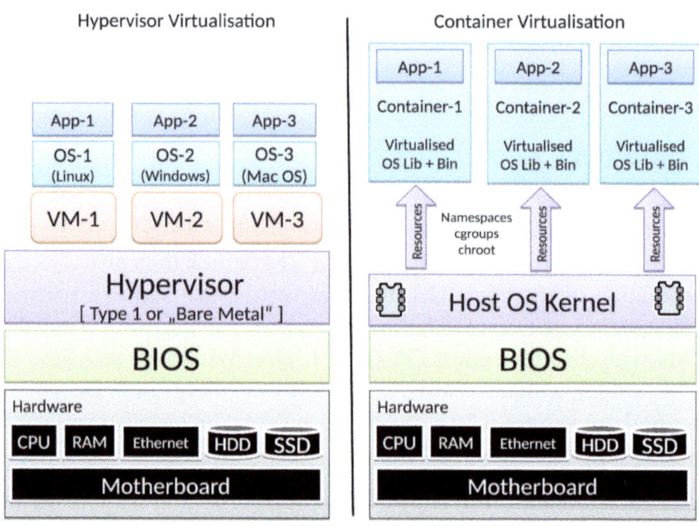

Fig. 14.5 Comparison of virtual machine and container virtualistation

The container technology uses three key features of the Linux kernel, namespaces, cgroups, and chroot which respectively allow to isolate processes, fully manage resources, and ensure an appropriate security [3]. Namespaces allow a Linux system to abstract a global system resource and make it appear to the process inside a namespace like an own isolated instance of the global resource. In other words, namespaces can be understood as a means to provide processes with a virtual view of their operating environment. The term *cgroups*[6] is an abbreviation for "control groups" and allows to allocate resources like CPU, memory, disk space, and network bandwidth to each container. The chroot feature can isolate namespaces from the rest of the system, and protect against attacks or interferences from other containers on the same host.

An important difference between a Linux kernel running several containers or several processes (in the classical multi-tasking concept) is the ability to isolate, as code running on one container cannot accidentally access resources or interfere with processes running in another container through the kernel. Another important feature or concept of container systems is that they are immutable (unchangeable). This means that in case a container based application code is modified, no (physical or virtual) machine needs to be taken down, updated, and rebooted again. Instead, a new container image with the new code is generated and "pushed" out to the cluster. This is an important aspect that supports modern software development

---

[6] cgroups was originally developed by Google engineers Paul Menage and Rohit Seth under the name "process containers" and mainlined into the Linux kernel 2.6.24 in 2008 [5].

processes that build on continuous integration and delivery (CI/CD) and continuous deployment as main corner stones.

One widely used software suite to build, manage, and distribute container based software is Docker® [6], which is also the basis for open industry standards like the container runtime specification (runtime-spec [7]) and the container image specification (image-spec [8]) as defined by the Open Container Initiative [9].

14.3 Orchestration: Kubernetes®

Containers are today a very important tool for the development and operations of cloud-native applications[7] that mainly build on a *microservices* architectural approach. This defines the functions of an application as a delivered service which can be built and deployed independently from other functions or services. This also means that individual services can operate (and fail) without having a negative impact on others. Such architectures can require the deployment of hundreds or even thousands of containers which makes their management challenging. For this purpose, *container orchestration* tools were invented which support the following functions: [10]

- container provision and deployment,
- container configuration and scheduling,
- availability management and resource allocation (e.g., fitting containers onto nodes),
- container scaling (i.e., automated rollouts and rollbacks) based on balancing workloads,
- network load balancing and traffic routing,
- container health monitoring and failover management (self-healing),
- application configuration based on the container in which they will run,
- ensuring secure container interactions.

There are several container orchestration tools that can be used for container lifecycle management, some popular options are Docker® in swarm mode [11], Apache Meso® [12], or Kubernetes® (also referred to as K8s®) [13].

Kubernetes is an open source container orchestration tool that was originally developed and designed by engineers at Google and donated to the Cloud Native Computing Foundation in 2016 [14]. The Kubernetes software consists of the following main components (see also Fig. 14.6) [15]:

- Nodes (Minions) are machines (either physical or VMs) on which the Kubernetes software is installed. A node is a worker machine (bare-metal server, on-premises

---

[7] This term refers to software applications that are designed and tested in a way to allow simple deployment across private, public, and hybrid clouds.

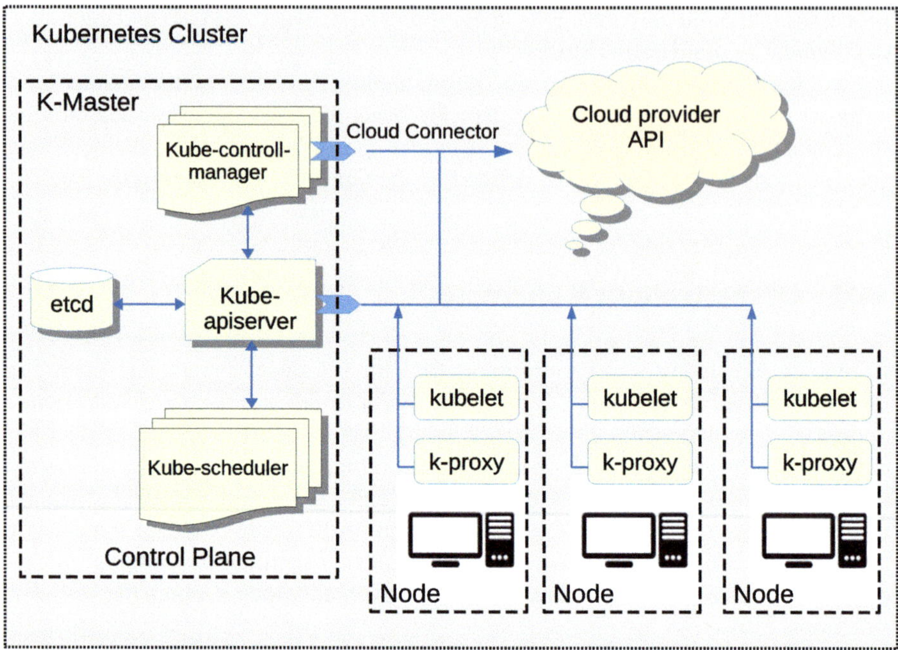

Fig. 14.6 The main components of a Kubernetes® cluster (refer to text for more detailed description)

VM, or VM on a cloud provider) which hosts a container runtime (for example a Docker Engine) to launch containers inside a pod (see pod definition below).

- A cluster defines a set of nodes grouped together to allow automatic failover. If one node fails, the applications running on it will be launched from the other node (done by *kubelet*). Clusters are also used to improve the computational and networking load balancing (done by *kube-proxy*). A cluster also contains one or more master nodes that run the Kubernetes *control plane* which consists of different processes, such as an API server (allowing applications to communicate with one another), scheduler (selects nodes to run containers based on their resource requirements), controller manager (runs controllers for nodes availability, replication, endpoints, etc.), and *etcd* (a globally available configuration store).

- A pod is the smallest deployable unit that can be managed by Kubernetes. A pod is a logical group of one or more containers that share the same IP address and port space. The main purpose of a pod is to support co-located processes, such as an application server and its local cache. Containers within a pod can find each other via localhost and can also communicate with each other using standard inter-process communications like SystemV, semaphores, or POSIX shared memory. In other words, a pod represents a "logical host".

References

1. VMware Incorporation. (2022). https://www.vmware.com/products/vsphere.html, Accessed Mar 21, 2022.
2. Rosen, R. (2016). Namespaces and cgroups, the basis of Linux containers. In *Proceedings of NetDev 1.1: The technical conference on linux networking (February 10–12, 2016. Seville, Spain)*. https://legacy.netdevconf.info/1.1/proceedings/slides/rosen-namespaces-cgroups-lxc.pdf, Accessed Apr 9, 2022.
3. Cloud Academy. (2022). *Container virtualisation: What makes it work so well?* https://cloudacademy.com/blog/container-virtualization/, Accessed Mar 24, 2022.
4. IBM. (2022). *Containers vs. virtual machines (VMs): What's the difference.* https://www.ibm.com/cloud/blog/containers-vs-vms, Accessed Mar 24, 2022.
5. Corbet, J. (2007). *Process containers LWN.net.* https://lwn.net/Articles/236038/ Accessed Apr 9, 2022.
6. Docker Inc. (2022). https://www.docker.com/, Accessed Mar 24, 2022.
7. Open Container Initiative. (2022). *OCI runtime-spec v. 1.0.2 release notice.* https://opencontainers.org/release-notices/v1-0-2-runtime-spec/, Accessed Apr 10, 2022.
8. Open Container Initiative. (2022). *OCI image-spec v. 1.0.2 release notice.* https://opencontainers.org/release-notices/v1-0-2-image-spec/, Accessed Apr 10, 2022.
9. The Open Container Initiative. (2022). About the open container initiative. https://opencontainers.org/about/overview/, Accessed Apr 10, 2022.
10. Red Hat. (2019). *What is container orchestration?* https://www.redhat.com/en/topics/containers/what-is-container-orchestration, Accessed Apr 10, 2022.
11. Docker Incorporation. (2022). *Swarm mode overview.* https://docs.docker.com/engine/swarm/, Accessed Apr 10, 2022.
12. Hindman, B., Konwinski, A., et al. (2011). Mesos: A platform for fine-grained resource sharing in the data center. In *Proceedings of the 8th USENIX symposium on networked systems design and implementation (NSDI 11)*. https://www.usenix.org/conference/nsdi11/mesos-platform-fine-grained-resource-sharing-data-center, Accessed Apr, 10 2022.
13. Kubernetes. (2022). https://kubernetes.io/, Accessed Apr 10, 2022.
14. Cloud Native Computing Foundation. (2022). *Projects kubernetes.* https://www.cncf.io/projects/kubernetes/, Accessed Apr 10, 2022.
15. Gupta, G. (2019). *Components of kubernetes architecture.* https://kumargaurav1247.medium.com/components-of-kubernetes-architecture-6feea4d5c712.

Chapter 15
Operations

The most common picture we associate with spacecraft operations are engineers in a control room sitting in front of a console as shown in Fig. 15.1 where they analyse new telemetry received from orbit, prepare telecommand sequences for future uplink, or talk to astronauts on the voice loop in case of a manned mission. Whereas these activities are for sure the most prominent tasks in satellite operations, there are also many others which are less noticeable but still play an important role.

The aim of this chapter is to provide an overview of all the main operational tasks and responsibilities of ground operations engineers. These can be grouped into three main categories according to the project phase they are relevant for. The first set of activities are described in Sect. 15.1 and need to be performed during the early project phase when the ground and space segment requirements need to be defined. The second category is relevant during the integration, verification, and validation of the ground and space segments (see Sect. 15.2), and the third one after launch, once in-orbit flight operation takes place (see Sect. 15.3).

As the overall operations concept is usually quite project specific and will also depend on the type and background of the operator,[1] the operational processes and corresponding terminology might also vary significantly. It is therefore worth to consult existing standards like e.g., ECSS-E-ST-70C [1] titled *Ground Systems & Operations* which can be easily tailored to the applicable project size and needs. The terminology and description presented here also follows this standard.

---

[1] The term *operator* in this context refers to the entity that is contractually responsible to provide all operations task as a service to its customer. A more precise term would therefore be *operations service provider*, which could be a public (space) agency or a commercial company.

B. Nejad, *Introduction to Satellite Ground Segment Systems Engineering*, Space Technology Library 41, https://doi.org/10.1007/978-3-031-15900-8_15

Fig. 15.1 View of the flight directors console in the Houston Mission Control Center (Johnson Space Center) during the Gemini V flight on August 21, 1965. Seated at the console are Eugene F. Kranz (foreground) and Dr. Christopher C. Kraft Jr. (background). Standing in front of the console are Dr. Charles Berry (left), an unidentified man in the centre, and astronaut Elliot M. See. Image Credit: NASA [2]

15.1 Preparation and Definition Phase

15.1.1 Requirements

In the early phases of a space project the main focus is the definition and consolidation of the mission requirements which translate the high level needs and expectations of the end user or customer into mission requirements (see also Chap. 2). These are documented in the *project requirements* that are the basis for the derivation of lower level requirements for the space and ground segments. An important activity for the operator at this stage is to review these requirements and assess the operational feasibility of the mission, as a project with an unrealistic or even impossible operational concept will never be successful. Furthermore, this review needs to encompass key design documents of the space and ground segment like the satellite user manual, the detailed design definition files, performance budget files, and risk registers. Furthermore, the operator needs to support the definition of ground segment internal and external interfaces to ensure that all information required to operate the satellite is properly defined and provided at the times needed. The operator needs to document the specific needs and a detailed description on how the mission will and can be operated in the *Mission Operations Concept Document* (MOCD) which also serves as an additional source to derive design requirements for the space and ground segments. Once the requirements baseline has been established, the operator should participate in the requirements review milestone (refer to SRR in Chap. 2) and focus on the following operational key areas in the ground control segment:

- Functionality that allows to evaluate the satellite performance and investigate anomalies based on the analysis of all received telemetry.
- Requirements specific to algorithms and tools needed for mission planning and contact scheduling. Special attention should be given to efficiency and performance requirements (e.g., efficient use of ground station resources, consideration of all relevant planning constraints, planning of backup contacts, generation time of mission plans and schedules).
- Completeness in the specification of flight dynamic related needs, especially the ones that go beyond the standard routine functions (e.g., advanced mission analysis tools, manoeuvre planning optimisers, launch window analysis, etc.).
- Specification of the ground segment automation functionality for satellite monitoring and commanding. The defined concept for automated contact handling must be able to support the operations concept and consider the foreseen flight control team staffing profiles (e.g., FCT main and shift team sizing). In case of incompatibilities, additional FCT staffing might be needed to operate the mission which might generate additional costs that exceed the foreseen budget in the mission concept and return of investment plans.
- Ensure the correct specification of any functionality required to operate and maintain all remote assets that cannot be easily accessed by operators and maintainers. A typical example would be an unmanned remote TT&C station that is located an a different continent.

15.1.2 Mission Analysis

A further important contribution of the operator during this phase is to perform mission analysis which is required for the proper choice of the operational orbit geometry and the corresponding sizing of spacecraft subsystems. The orbit geometry must be driven by the needs and constraints of the payload and should allow an optimum sensor performance or service provision to the end user. The following orbit geometries are usually considered and need to be further refined to meet the specific project needs:

- Low altitude (LEO) and high inclination (polar) orbits which provide access to a large portion of the Earth's surface and are therefore the prime choice for Earth observation payloads. Subcategories are the Earth-synchronous orbit with a defined ground-track repeat cycle (and revisit periods of certain target areas) and the Sun- synchronous orbit which provides a constant orientation of the inertial orbit plane with respect to the Earth-Sun vector.

- Medium altitude orbits (MEO) are usually used by GNSS type satellites as they provide favourable conditions for navigation receivers on ground (e.g., the simultaneous visibility of a minimum of four satellites to achieve a position fix).
- Geo-synchronous orbits (GEO) provide a constant sub-satellite point on the Earth's surface which is required by broadcasting (TV or telecommunication) or weather satellites.
- Interplanetary trajectories are quite complex and often use gravity assist from planets to reach a target object. The definition of such trajectories are very project specific and require an extensive amount of effort for their design and optimisation.

Another aspect that needs to be analysed is the TT&C ground station network which needs to be properly dimensioned with respect to the number of antennas and their geographic distribution. Both will impact the possible contact frequency and the contact durations which themselves determine the data volume that can be downlinked (e.g., per orbit revolution). This must be compatible with the data volume that is accumulated by the satellite subsystems and the capacity of onboard storage that can be used to buffer telemetry onboard.

The geographic distribution of the antennas in a ground station network impacts the achievable orbit determination accuracy which is derived from the radiometric measurements (see Chap. 6). The required size of the ground station network will usually differ for a LEOP phase and during routine operations and both cases need to be considered separately.

The mission analysis should also comprise the estimation of maximum eclipse durations that need to be supported by the satellite. This determines the required size of the solar panels and the battery storage capacity. Environmental impacts like the expected accumulated radiation dose determines requirements for additional satellite shielding or the use of radiation hardened electronic components.

The launch window and orbit injection analysis is used to select the correct launch site, epoch, and azimuth. The estimation of the overall Δv budget needed for the entire mission is required to properly size the fuel tanks. The budget needs to consider the needs to reach the operational target orbit during LEOP and also maintain it throughout the entire projected service time according to the station keeping requirements. Additional margins need to be considered that allow to de-orbit the satellite or place it into a disposal orbit at the end of its lifetime.

15.2 Procedure Validation and Training

Once the space and ground segment have passed their design phase and reached their implementation stage, the operator must shift focus to the development of operational products which need to be ready (and validated) well in advance of the planned launch date. A very important product is the complete set of *flight operations procedures* (FOPs) that describe in detail the steps to be performed by

<table>
<tr><td colspan="7">Satellite Flight Operations Procedure: FOP-AOCS-0010
OOP Update – Version 1.2 (10/10/2021)</td></tr>
<tr><td colspan="7">Objective: This procedure is used to update the on-board propagators. This procedure updates the parameters of the STS module: Satellite, Sun and Moon orbits at specific epoch, not necessarily the three at once. In fact, satellite, Sun and Moon orbit updates shall be performed at different frequencies</td></tr>
<tr><td colspan="7">Constraints: Flight Dynamics input available (SCOS2K TPF transferred to MCS)</td></tr>
<tr><td colspan="7">Satellite configuration: SAT-NOM/AOCS-NOM/EPS-NOM
OBDH software is running correctly, TM packets activated: SAA_TMCSTS (SPID 258612)</td></tr>
<tr><td>Step</td><td>Label</td><td>Activity/Remarks</td><td>Telecommand</td><td>Telemetry</td><td colspan="2">Display/Branch</td></tr>
<tr><td>1</td><td></td><td>Initial Checks</td><td></td><td></td><td colspan="2"></td></tr>
<tr><td>1.1</td><td></td><td>Check Packets Enabled</td><td></td><td>SAA_TMCSTS</td><td colspan="2">TM-AOCS-1</td></tr>
<tr><td>1.2</td><td></td><td>Check OOP Telemetry (SC)</td><td></td><td></td><td colspan="2"></td></tr>
<tr><td></td><td></td><td>Verify Telemetry X in ECI</td><td></td><td>AAAT076D</td><td colspan="2"></td></tr>
<tr><td></td><td></td><td>Verify Telemetry Y in ECI</td><td></td><td></td><td colspan="2"></td></tr>
<tr><td></td><td></td><td>Verify Telemetry Z in ECI</td><td></td><td></td><td colspan="2"></td></tr>
<tr><td>2</td><td></td><td>Command OOP update</td><td></td><td></td><td colspan="2">TC-AOCS-1</td></tr>
<tr><td>2.1</td><td></td><td>Execute Telecommand SAASTS00 SET STS
AAAP123H = SAT_TIME, AAAP123D = SAT_SMA
AAAP123A = SAT_INC
AAAP123A = SAT_RAAN
AAAP123A = SAT_AOP
AAAP123_ = SAT_ECC
AAAP123V = DSAT_SMA
AAAP123V = DSAT_INC</td><td>SAASTS00</td><td></td><td colspan="2"></td></tr>
<tr><td>3</td><td></td><td>Verify OOP Update</td><td></td><td></td><td colspan="2"></td></tr>
<tr><td>3.1</td><td></td><td>Verify OOP Telemetry (ECI Cartesian)
AAAT0125 = SATPOSXECI</td><td></td><td>AAAT0125</td><td colspan="2">TM-AOCS-2</td></tr>
<tr><td></td><td></td><td>AAAT0125 = SATPOSYECI</td><td></td><td>AAAT0126</td><td colspan="2">TM-AOCS-2</td></tr>
<tr><td></td><td></td><td>AAAT0125 = SATPOSZECI</td><td></td><td>AAAT0127</td><td colspan="2">TM-AOCS-2</td></tr>
</table>

Fig. 15.2 Example for the format and contents of a very simplified flight operations procedure (FOP) for the update of the onboard orbit propagator

ground operators in order to complete a specific satellite task. These procedures also provide a detailed description of key activities, their prerequisites or conditions[2] and "GO/NOGO" criteria for critical commands.

In human space flight where astronauts are involved and can interact with the ground crew, such FOPs could in theory also be in free text format. In unmanned missions they will usually have a clearly defined tabular format and are generated using specialised software products that host the satellite TM/TC database for cross referencing and consistency checks.[3] Well written FOPs should as a minimum contain the following inputs (see also example shown in Fig. 15.2):

- The name of the procedure, a reference or short acronym, the version number, and a change record.
- A detailed description of the objective that is supposed to be achieved.
- A list of all conditions and constraints that exist as a prerequisite for its execution.
- The required satellite configuration at the start of the procedure (e.g., which system mode, attitude mode, payload mode, etc.).
- A tabular list with the detailed steps including the exact references to telemetry parameters (mnemonic) and telecommands (see also recommendations in [3]).

---

[2] Examples could be specific eclipse conditions, Sun or Moon presence in a sensor FOV, or the visibility of the satellite by a ground station.

[3] See also the description of the *operations procedure editor* in Chap. 9.

FOPs are usually grouped into different categories which is also reflected in their short reference or acronym. The grouping can either consider the level of satellite activity (e.g., system procedure versus subsystem procedures) or the mission phase they are used for (e.g.,LEOP, routine, or contingency operations). This categorisation also helps in organising the operator training and allows the formation of specialised sub teams that are dedicated to a specific type of activity. As an example, a shift team that is meant to perform overnight and weekend satellite monitoring must not necessarily be familiar with the more complex special operations tasks required during a LEOP or a special operations phase.

The generation of FOPs requires a very detailed knowledge of the satellite and its avionics software. An initial set should therefore always come directly from the satellite manufacturer who knows the spacecraft very well and can therefore best describe the required satellite configuration and applicable constraints to perform a satellite activity. Also additional information and recommendations that might be relevant for commanding can be added. The operator has to review and validate this initial set of procedures and potentially even extend them with ground segment specific information that was not known to the satellite manufacturer.

The FOP validation must be performed on a representative ground control segment (VAL chain) which is either connected to a satellite simulator (running the correct version of OBSW as described in Chap. 11) or to the actual satellite, if still located at its AIT site or a test centre. The detailed scope of the validation campaign must be clearly defined and documented as part of an operational validation plan, which will ensure that all FOPs (and their correct versions) have been properly validated and are ready to be used in flight.

The development and validation of *ground operations procedures* or GOPs is another important task the operator has to perform at this stage. These procedures define in detail all the required steps at ground segment level in order to generate, import, or export a specific product. Operational GOPs are usually derived from the software operations manuals provided by the element manufacturers and need to be tailored to the specific operational needs. Examples of such operational GOPs could be the generation of an event file, a set of manoeuvre command parameters, or a mission plan. A different type of GOPs are the maintenance procedures which describe the detailed steps to keep the ground segment configuration up to date. Examples here could be the update of an anti-virus definition file, the alignment of a leap second offset value required for correct time conversion (see Appendix B), or even hardware specific tasks like the replacement of a faulty hard disk, RAID controller, or even an entire server. A specialised subset are the *preventive* maintenance procedures which define and detail specific checks or activities to be performed at a regular basis in order to reduce the wear-out of certain components in the segment and prolong the lifetime of hardware (e.g., check of cooling fans on server racks or the correct level of grease inside the gear boxes of a TT&C antenna system).

The formation and training of all operations staff needs to be done during this phase in order to be ready for the day of launch. The following team structure is usually adapted (see Fig. 15.3)

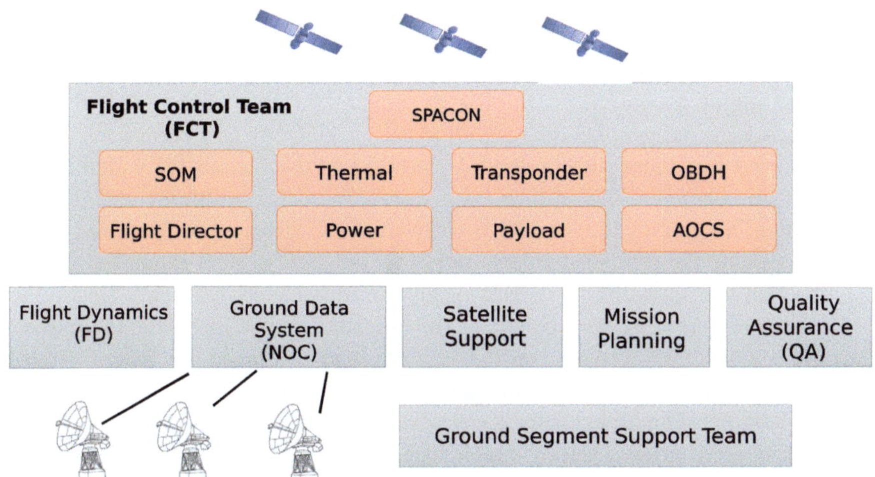

Fig. 15.3 Possible structure of a spacecraft operations teams according to ECSS-E-ST-70C [1, 3]. SOM = Spacecraft Operations Manager, OBDH = Onboard Data Handling, AOCS = Attitude and Orbit Control System

- The mission or *flight control team* (FCT) is in charge of the overall control of the space segment and is headed by the *spacecraft operations manager* (SOM) and the *flight director* (FD). All major subsystems of the spacecraft should be represented as dedicated console positions in the control room and must therefore be manned by a subsystem specialist. Having dedicated operators to monitor each of the subsystem will lower the risk to overlook an out-of-limit telemetry value and also ensures that the adequate level of expertise is present, in case an anomaly occurs that requires a deeper understanding of a specific subject matter. Another important role in the FCT is the *command operator* or *spacecraft controller* (SPACON) who is in charge to operate the Spacecraft Control Facility (see Chap. 7) in order to load and release all the TC sequences to the satellite. It is important that only one role in the entire FCT is allowed to perform this highly critical task as this will avoid any uncontrolled or contradictory TC transmission to the satellite.
- The *flight dynamics* (FD) team is responsible to perform all relevant activities to maintain an up-to-date knowledge of the orbit and attitude position and to generate all orbit related products (e.g., event files, antenna pointing, orbit predictions etc.). The FD team is also responsible to generate the relevant telecommand parameters for manoeuvres or orbit updates (OOP) that need to be provided to the FCT for transmission to the satellite.
- The *ground data systems* (GDS) team has to coordinate the booking and scheduling of TT&C stations. The infrastructure provided for this is referred as the *network operations centre* or NOC and is of special importance during LEOP phases when additional ground station time has to be rented from external

networks in order to provide additional contact times to support critical satellite activities.

- The *ground segment support team* is in charge of the maintenance of the entire ground segment equipment. Similar to the organisation of the FCT, this team is headed by the *ground operations manager* (GOM) and comprises a set of specialised engineers with in-depth knowledge of the various GCS elements and the entire segment. This allows them to resolve daily problems and if needed document anomalies that require a software fix. Software patches can be either developed by the ground segment support team (in house) or requested from the entity that is responsible to delivery regular maintenance releases.[4]
- The *mission planning* team generates the operational timelines which requires the coordination of all activities on the satellite but also inside the ground segment (e.g., maintenance activities, software and hardware replacements, etc.). Another important planning aspect is the sharing and scheduling of equipment and human resources (e.g., ground station contact time, on console shifts, support outside of working hours).
- The *satellite support* team is composed of experts from the satellite manufacturer and is responsible to provide technical support in case of anomalies or contingency situations that occur on the satellite in orbit. This team is usually present in the control centre during times of critical operation (e.g.,LEOP, complex manoeuvre sequences, or safe mode recoveries).

15.3 In-Flight Phase

The in-flight phase can be subdivided into the Launch and Early Orbit Phase (LEOP), the commissioning, the routine, and the disposal sub-phases. LEOP is by far the most demanding undertaking, as it requires a number of highly complex and risky activities to be performed as depicted in the timeline in Fig. 15.4. It starts with the launch vehicle lifting off from a space port, going straight into the ascent phase during which a launch vehicle dependant number of stages are separated after their burn-out. The last one is the *upper stage* which has to deliver the satellite into its *injection* orbit. Once arrived, the satellite is separated from the launch dispenser and activates its RF transponder. The first critical task for the operations team is to establish the very first contact from ground using the pointing information based on the expected injection location. In case of a launch anomaly, the actual injection point might however deviate from the expected one which could imply an inaccurate antenna pointing making the first contact to fail. In this worst case scenario, a search campaign with several ground antennas must be performed. If this is not successful, passive radar tracking might even be needed. Once the ground station link could

---

[4] In case an urgent software fix is needed, a special process for the development and installation of an emergency software patch should be exercised.

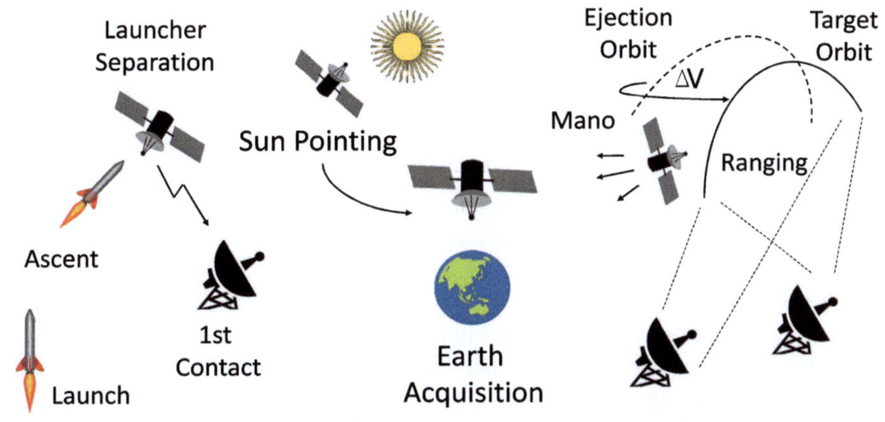

Fig. 15.4 Important activities during the Launch and Early Orbit Phase (LEOP)

be established, the first radiometric data can be gathered and used for a first orbit determination to get more accurate pointing vectors for subsequent contacts. From the first received housekeeping telemetry dump an early initial check up of all major satellite subsystems can be performed which allows to determine whether they have survived the launch and are operating nominally.

As the stability of the radio link also depends on the satellite orientation, it is important to gain a stable attitude as soon as possible. During the launch phase and right after injection the satellite depends on battery power with only limited capacity. The immediate deployment of the solar panels and their proper orientation to the Sun is therefore one of the first tasks to be performed. A favourable first orientation is the so called Sun pointing attitude in which the body axis that is perpendicular to the solar panel is pointed into the direction of the Sun. For stability reasons, the entire satellite body is then rotated around that Sun pointing axis.

After the batteries have been sufficiently charged and the first check ups of the main subsystems finalised, the satellite can be transitioned into an Earth pointing orientation.[5] As the satellite is usually not directly injected into its target orbit,[6] a set of orbit change manoeuvres must be performed in order to reach its operational orbit. Accurate orbit correction requires radiometric tracking of the satellite before and after each orbit manoeuvre in order to estimate the burn execution error and obtain an up-to-date orbit vector. This requires the regular scheduling of ranging

---

[5] This is of course only applicable for satellites with a payload that requires to be oriented to the Earth surface, which is not the case for interplanetary spacecraft or space telescopes.

[6] This allows to keep sufficient distance between the operational orbit and the launcher stage, mitigating the risk of collision.

campaigns during the manoeuvre phase which demands more frequent contacts from the ground station TT&C network.

The LEOP phase ends once the satellite has reached its designated orbit and performs in a stable configuration. The next step is the commissioning phase, during which a more thorough in-flight validation of both platform and payload is performed. Depending on the complexity of this activity, some projects might even distinguish between platform and payload commissioning phases. One important aspect to be tested is the proper functioning of component redundancies, as satellite manufacturers usually build in hot or cold redundancies for critical subsystems, where hot-redundant devices have a backup component that continuously runs in parallel to allow a very fast hand-over in case of failure. The cold redundant devices in contrast need to be powered on and "booted" if the prime fails which implies a certain latency during the switching phase. The testing of cold-redundant devices is more risky as the state of the device is not known and could be deteriorated, after having been exposed to the harsh launch environment. Despite this risk, the testing is important in order to gain trust in the full availability of all redundancies which might be required at a later stage in the mission. Payload commissioning can also comprise calibration campaigns of complex instruments or the accurate alignment of the main antenna beam which might require additional expert support and specialised ground equipment during that phase.[7]

Once the commissioning phase is finished, the routine phase starts which is usually the longest phase of the entire satellite lifetime. During this phase the satellite and its payload have to provide their service to the end user. For commercial projects this is also the time to reach the break even point or return of investment and achieve its projected profit. Depending on the mission profile, routine contacts are usually planned once per orbit in order to dump the house-keeping telemetry, uplink all required telecommands, and perform ranging measurements. The routine phase can be supported by a much smaller operations team with emphasis on long term monitoring and trend analysis of the most relevant subsystem telemetry. Also regular orbit determination needs to be performed by the flight dynamics team in order to monitor the satellite's deviation from the reference trajectory and perform station keeping manoeuvres to correct it. Special manoeuvres could also be required in case the satellite risks to collide with another space object.

The end-of-life of a satellite can be either determined by the depletion of its fuel tank or a major malfunctioning of a key subsystem that cannot be replaced by any onboard redundancy. Either of these events implies that the service to the end user cannot be maintained anymore and the satellite needs to be taken out of service. This marks the start of the disposal phase which should ensure that the satellite is either removed from space[8] or transferred into a disposal orbit where it does not disturb or cause any danger for other satellites in service. For geostationary satellites, the

---

[7] This sub phase is also referred to as *in-orbit testing* or IOT.

[8] Satellite de-orbiting is mainly an option for orbit altitudes low enough to be exposed to sufficient atmospheric drag.

removal of the old satellite from the operational slot is also commercially relevant in order to make space for its replacement allowing service continuity. For satellites that must remain in orbit, the depletion of all remaining fuel in the tanks or any latent energy reservoirs is very important and today even mandatory [4]. This satellite passivation avoids the risk of a later break up or explosion that would contribute to a further increase of the space debris population, which already poses a major concern for safe satellite operations today. The chosen disposal orbit must also fulfil adequate stability requirements that ensure it does not evolve into an orbit that could cause a collision risk for operational satellites in the near and even far future.

References

1. European Cooperation for Space Standardization. (2008). *Space engineering, ground systems and operations*. ECSS. ECSS-E-ST-70C.
2. National Aeronautics and Space Administration. (2019). In S. Loff (Ed.) Image source: https://www.nasa.gov/image-feature/kranz-on-console-during-gemini-v-flight, last updated Jul 22, 2019.
3. Uhlig, T., Sellmaier, F., & Schmidhuber, M. (2015). *Spacecraft operations*. Springer. https://doi.org/10.1007/978-3-7091-1803-0_4.
4. United Nations Office for Outer Space Affairs. (2010). *Space debris mitigation guidelines of the committee on the peaceful uses of outer space*. https://www.unoosa.org/pdf/publications/st_space_49E.pdf Accessed Mar 23, 2022.

Chapter 16
Cyber Security

Cyber crime has significantly increased in the past decade and is today considered as a major threat for private IT users, companies, and even more for critical public infrastructure like power plants, electric distribution networks, hospitals, airports, traffic management systems, or military facilities. Cyber attacks are a very peculiar kind of threat because they require only very basic IT equipment and a persons' skills, but can still cause a significant amount of damage even to a large and complex infrastructure. Attacks can be launched any time by an individual or a group of people, independent of age and location. Even if an attack has already been launched, it might take time before the victim realises the infiltration, and meanwhile the target computer could be transformed into a "zomby" machine that itself initiates an attack. In such cases, it might even be extremely difficult to backtrace the real aggressor and make the right person responsible for the caused damage. To better understand potential sources for cyber threats and the corresponding risk, it is worth to explore the motivation of a cyber attack, which can fall into one or even a combination of the following categories:

- Monetary driven attacks aim to gain access to bank accounts or credit cards and perform illegal transactions to a rogue account. Alternatively, an attack can encrypt data on the hard disk of a target host with the initiator subsequently black mailing the victim to make a payment to regain access to the data (ransomware). In both cases, the dominant motivation is financial profit.
- Espionage driven attacks in contrast try to remain fully invisible to the victim and remain undiscovered for as along as possible. During this time, all kind of sensitive data are illegally transferred from the infected host to the attacker's machine. Potential target data can be intellectual property of a company (industrial espionage) or military secrets. The infiltrated machine could also be used as an eavesdropping device to transfer keystrokes or audio and video recordings via a connected webcam.
- Sabotage driven attacks aim to disturb or even destroy the operational functionality of an application, a target host, or even an entire facility. Especially industrial

B. Nejad, *Introduction to Satellite Ground Segment Systems Engineering*, Space Technology Library 41, https://doi.org/10.1007/978-3-031-15900-8_16

systems such as refineries, trains, clean and waste water systems, electricity plants, and factories which usually are monitored and controlled by so called Supervisory Control and Data Acquisition (SCADA) systems[1] can be likely targets for this kind of attack. Well known examples are the 2011 Stuxnet cyber attack on Iran's nuclear fuel enrichment plant in Natanz or the 2015 attack on the Ukrainian power grid [2].

Even if one might argue that the motivation of an attack might be of secondary relevance, it is still worth to be considered as it might point to the person or entity who has launched it. Monetary driven cyber crime will usually be committed by a single entity called *black hat hacker*,[2] who's primary target are privately owned IT systems which are usually less protected and their owners are more susceptible to hand out sensitive credit card information (e.g., responding to phishing emails) or a pay the hackers' ransom in order to regain data access. Industrial or military espionage and sabotage driven attacks are more likely to be launched by governments and their respective intelligence agencies who are interested to acquire a very specific type of information that is of high strategic value to them. These entities are therefore ready (and able) to invest a considerable amount of effort (and money) to carefully design, plan, and launch a highly complex orchestrated attack that could exploit not only one but even an entire range of zero-days vulnerabilities in order to be successful.

The ground control segment of every space project represents a complex IT infrastructure that comprises a substantial amount of servers, workstations, operating systems, application software packages, and interface protocols in a single location. Especially for projects with high public profile and strategic relevance (both for civilian and military purposes), the espionage and sabotage driven attacks must be considered as a likely threat scenario. An appropriate level of cyber hardening therefore needs to be an integral part of the ground segment design and has to also be considered in the maintenance plan throughout the entire operational lifetime.

16.1 Attack Vectors

The detailed methodology and structure of a cyber attack can be very complex, especially if targeted to complex and well protected systems. It is therefore not possible to provide a generic and at the same time accurate description of the

---

[1] SCADA systems are based on sensors and actuators that interact in a hierarchical way with field control units located either remotely and/or in inhospitable environments. They usually use Programmable Logic Controllers (PLC) that communicate with a so called fieldbus application protocols like Modbus, which in current versions have only few security features implemented [1].

[2] This term is used in combination with *white hat hackers*, representing computer specialists that exercise hacking with the aim to discover vulnerabilities of a system in order to increase its security.

different attack strategies. It is however a valid assumption to state that even a complex attack scenario will build on a basic set of standard methods which are referred to as *attack vectors*. The examples provided in this section are a basic set of such vectors and should by no means be considered as exhaustive or up-to-date. The aim is to help the reader to gain a basic understanding and awareness of the nature and techniques applied by such vectors.

16.1.1 Password Hacking

Password hacking is one of the most efficient methods to gain access to a system's data and its application. A system usually stores all registered passwords in encrypted format generated with a *hash* function. A hash algorithm maps (input) data of arbitrary size into a fixed length hash code, which is supposed to be collision resistant. This means that there is a very low probability to generate the same hash code from two different input values. The hashing function only works in one direction (i.e., clear text into encrypted format) which makes it impossible to reconstruct the input data from the generated hash code. The only way to unveil hashed passwords is therefor by guessing a large amount of input values and compare their hashed outputs to the saved values on the system. This can either be done using a trial and error technique, referred to as a *brute force attack*, or with more elaborated techniques that use password dictionaries or rainbow tables. An important countermeasure to password hacking is the increase of password complexity and a policy that enforces regular password modifications.

16.1.2 Back Door Attack

Back door attacks are designed to exploit either existing or newly created entry points of a computer hardware, operating system, or application software. The term *backdoor* in this context describes any means to access a system, bypassing the nominal authentication process. Existing backdoors in a system could have been left by vendors for maintenance reasons which usually pose a low risk as they are widely known and therefore easy to secure (e.g., admin accounts with default passwords). More problematic are secret backdoors which are hard to detect as they could be embedded in proprietary software or firmware whose source code is not available to the public. Backdoors can also be actively created by malware like Trojans or worms which is use them to infiltrate the system with additional malware that could spy out passwords, steel sensitive data, or even encrypt data storage in order to ask for ransom (ransomware).

16.1.3   Distributed Denial of Service

The Distributed Denial of Service (DDoS) attack uses a large number of infected hosts in order to flood a target system with a very high number of requests that fill up either the available network bandwidth or the system's CPU or memory resources. The ultimate goal of such an attack is to make the target unresponsive for any legitimate user or service. Apart from the performance impact, there is also the risk for the target system to get infected with malware that turns it into an attack "handler" contributing to the overall DDoS attack (sometimes even without the knowledge of the system's owner).

16.1.4   Man-in-the-Middle

A *man-in-the-middle* (MITM) attack refers to a scenario in which the attacker redirects (relays) the nominal communication line between two parties without them being aware. A passive MITM attack is usually performed to simply eavesdrop on the conversation, whereas in an active MITM the attacker would himself relay messages (e.g., ask for specific information) making the recipient believe that the request comes from the expected other party. An efficient means to protect against MITM attacks is to always implement encrypted network connections for remote connections and interfaces using a *Virtual Private Networks* (VPN) configuration.

16.1.5   Root Access

The *root* account is an account with very high privileges in Unix and Linux based systems and is therefore a high risk target for illegal access. Hackers can use *rootkit* software packages that are designed to gain complete control over a target system or the network. It is therefore considered as a good practice to disable any root login accounts which are part of a default configuration of an OS when it is initially being deployed. An additional precaution is to adapt the *principle of least privilege*, which dictates that users should only be attributed the minimum set of rights necessary to perform the required work on the system.

16.1.6   Phishing

Phishing attacks refer to attacks that are designed to query sensitive information from a person (e.g., passwords or other personal identifiable information) without making the victim aware to hand that information to an unwanted source. This is often achieved by forging an email of a trustworthy entity (email spoofing) that requests the victim to access a link which actually redirects to a rogue website, but not the one the user thinks to connect to. Once the credentials have been handed out, they are immediately exploited.

16.1.7 Watering Hole

The *watering hole* and *drive by downloads* are attack vectors that infect legitimate websites with malicious scripts, which are designed to take advantage of specific vulnerabilities of the used web browser. Especially in the case of a drive-by download, malicious code is installed on the target computer. The infected websites are usually carefully chosen and based on the attacker's analysis of the user group frequently accessing it. This is also the reason for the choice of the name "watering hole", which makes reference to predators in nature who would rather wait for their prey near a watering hole rather than tracking it over long distances [3, 4].

16.1.8 Formjacking

Formjacking is a technique that bears some resemblance to the previously described watering hole attack, with the difference that formjacking is targeted to legitimate payment forms of banking and e-commerce payment portals. The malicious scripts injected into the websites capture and forward sensitive banking or other personal information to the attacker who will use this for various criminal purposes [4].

16.1.9 Malware

The illegal installation and execution of malicious software code on a computer system is the most frequent attack scenario and very often only one building block of a more complex attack vector. The generic term is *malware*, and more specific names are *computer virus, worms, trojans, spyware, adware*, or *ransomeware*. The most efficient ways to protect against malware is to control (and even reduce) the possibilities to import data (e.g., lock-down of external media ports and/or disconnection from WAN for system that don't require it), and to perform anti-malware scanning on a regular basis.

16.2   The Attack Surface

Cyber attacks exploit at least one but most of the time a number of so called *zero-day vulnerabilities*. These can be understood as design flaws that are present in a hardware item, operating system version, application, or a communication protocol which are not yet known to the vendor or programmer and can therefore be used for an attack. The time span between the discovery of a vulnerability and its fix by the vendor or programmer is called the *vulnerability window*. The *Common Vulnerabilities and Exposures* (CVE) numbering scheme provides a unique and common identifier for publicly known information-security vulnerabilities in publicly released software packages [5]. If a new vulnerability is unveiled, a CVE number is assigned by a so called CVE *Numbering Authority* or CNA which follows predefined rules and requirements on its numbering scheme and provides information on the affected products, the type of vulnerability, the root cause, and possible impact. A new record will be added to the CVE database which is publicly available (see official CVE search list [6]). The CNA has to make reasonable effort to notify the maintainer of the code so corrective measures can be taken. A fix for a CVE (or a number of CVEs) will usually be provided through the development and release of a software patch which is made available to the end users via the vendor's website.

The various areas of a system that are potentially susceptible to a cyber attack form the system's *attack surface* which is schematically depicted as the large sphere in Fig. 16.1. The arrows pointing from outside to the surface represent the

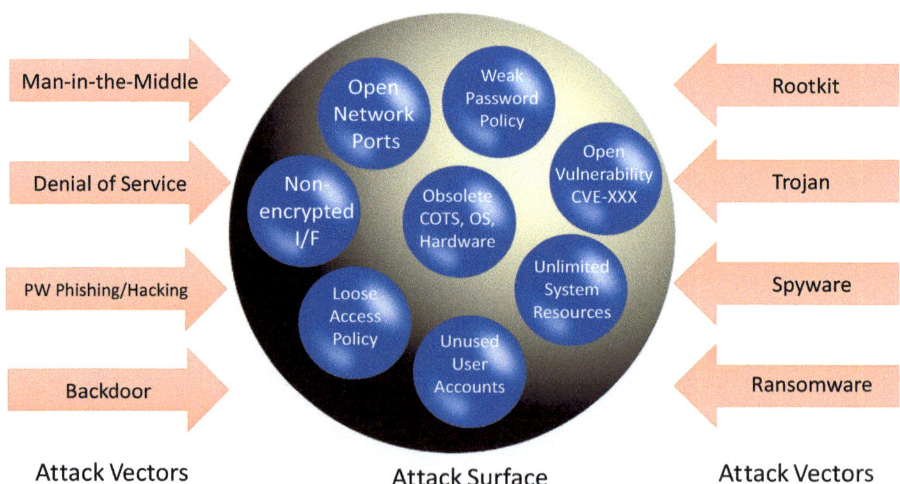

Fig. 16.1 Attack vectors (arrows) and attack surface (large sphere) of an IT system. The aim of cyber security is to reduce the attack surface by identifying and reducing the number of vulnerabilities represented by the small spheres

attack vectors which can breach and infiltrate the system and refer to the examples described in Sect. 16.1. The task of the cyber security officer is to continuously measure a systems' attack surface and try to reduce it as much as possible. A possible measurement metric is based on the identification of all system resources or weaknesses (shown as small spheres in Fig. 16.1), and to determine their respective contribution to the overall attack surface. The size of a single contribution can be understood as the product of the likelihood a specific resource could be exploited and the potential damage that could be caused to the system (refer to detailed explanation in [7]). In other words, a system with a large measured attack surface implies (1) a higher likelihood of an attack exploiting the existing vulnerabilities in the system (with less effort), and (2) the ability for an attacker to cause more significant damage to its target.

Looking at the typical ground segment architecture as described in this book, the following system resources are candidates for attacks and therefore need to be considered for the attack surface metric:[3]

16.2.1   Interface Protocols

The ground segment is composed of various elements or components that exchange information (signals) among them as per their defined interfaces. Interfaces should always be an area of concern, as the exchange of information requires the activation of network sockets (ports, IP addresses) and the use of standardised protocols (e.g., ftp, sftp, ssh, snmp, Remote Desktop Protocol or RDP, etc.). Especially the latter one can be a source for vulnerabilities, especially if older or deprecated versions of interface protocols are in use. The upgrade of interface protocols can impact many areas of the ground segment and therefore implies a significant effort when it comes to the re-qualification and non-regression testing. This might tempt a project to delay or postpone such an upgrade to not disturb ongoing operations and service provision ("never change a running system..."). This however needs to be carefully traded against the risk of an increase of the segment attack surface.

16.2.2   OS Vulnerabilities

The operating systems (OS) deployed on servers and clients should always be an area of attention and must therefore be continuously monitored for vulnerabilities and the corresponding updates or patches published by the vendor to fix critical cyber security issues. As such upgrades might be available quite frequently, a good

---

[3] The list provided here is not meant to provide a complete summary and the reader is encouraged to look out for more potential areas of concerns.

balance between their deployment and ongoing operational activities must be found. This must also comprise less visible IT components deployed at remote sites (e.g., the baseband modem or the antenna control unit of a TT&C station) as these devices might be subject to a looser access control policy and could therefore be more easily accessible for malware infiltration attempts.

16.2.3   Physical Port Access

Physical security measures aim to ensure the proper protection of physical hardware items like servers or workstations. Servers mounted inside racks are protected through the rack door lock mechanism, which is either operated with a physical key or a digital code (see also description in Sect. 13.4). This provides an easy means to control rack access and restrict it to authorised personnel. Furthermore, the access times can be monitored and logged. Locking devices for external media ports (e.g., USB, Ethernet, or optical ports) provide an additional level of security and further reduce the risk of malware infiltration through external media upload. This has an even higher relevance for workstations deployed in areas that are accessible to a much wider group of users where even additional measures like the logical lock-down of external USB ports and media devices are an essential measure.[4]

16.2.4   Network Port Security

Every operating system is configured to run a number of services which depend on a set of opened network ports at which they can listen to traffic. As any open port can in principle be abused by a cyber attack, it is highly recommended to restrict both the number of services and open ports (both TCP and UDP) to the number that is absolutely necessary. After the installation of a new operating system, the default port configuration needs to be scrutinised in detail, and a user or project specific lock-down applied. A software update might also require a temporary change of port configuration which has to be reverted. A port scan should therefore be performed at the end of every update activity in order to ensure that the required lock-down status of the entire system has been fully reestablished.

---

[4] A logical lock-down could foresee that only users with elevated rights (e.g., administrator or root) can access removable media such as USB storage or optical devices.

16.2.5 *Wake-on LAN*

Wake-on LAN is a technology introduced by Intel and IBM in 1997 [8] and allows to activate a dormant computer system that is connected to a (wired) network by sending a specially coded message called *magic packet*. This packet is sent to the network broadcast address on the data link layer (layer 2 in the OSI model [9]), and therefore reaches all devices on the same network. It contains an easily recognisable marker and a repetitive sequence of the computer's MAC address[5] which is intended to be activated. For this to work, the BIOS of the target computer's motherboard needs to have this capability enabled, which ensures that a part of the network adaptor remains active, even if the host machine is being powered off. Despite the good intention to give a system administrator an easy means to perform maintenance activities on remote machines (without having to physically visit them), this technology could be abused by anyone having gained access to the same LAN and become one building block in larger scope cyber attack strategy. It is therefore highly recommended to deactivate this feature, unless absolutely needed.

16.2.6 *Compilers and Interpreters*

Some operating systems come with installed compilers and/or interpreters for scripting languages. Examples are the GNU C++ compiler in certain Linux distributions or interpreters for shell, perl, or python scripts. While in a development environment the presence of compilers and interpreters makes absolute sense, they should not be needed on servers, workstations, or VMs that are used in a pure operational context. As compilers or interpreters can be used to create and execute malicious code, their presence on a system is a clear enabler for a cyber attack and should therefore be removed by default. In case interpreters are needed to execute operational scripts, their type, nature, and expected location must be clearly specified and documented.

16.2.7 *COTS Software*

Every ground segment uses application software that has been developed for the project to meet its specific requirements which is referred to as *bespoke* software. There will however always be some integration of COTS software from third party vendors, where COTS stands for *commercial off-the-shelf* and refers to any software product that can be procured on the open market and configured for the specific

---

[5] The Media Access Control Address or MAC address is a 48 bit alphanumeric number of an IT device able to connect to a network. It is also referred to Ethernet hardware address or physical address of a network device.

needs of the end user. As COTS software is widely used by many customers, it is also an attractive target for the exploitation of vulnerabilities through cyber attacks. A vulnerability in a COTS product could even be used as an entry point to infiltrate the bespoke software, which itself would not be an easy target due to its unknown architecture and code. All deployed COTS software packages therefore need to be listed and tracked for their version number and regular updates with available patches from the vendor performed. There might also be cases of very old COTS versions that are not maintained anymore but are still in use due to backwards compatibility with a bespoke software. If the replacement of such problematic packages is not immediately possible, alternative protective measures like a sandbox execution of an application should be envisaged.[6]

16.2.8   Remote File Access

Remote file access protocols are frequently used by client-server applications in order to better organise and centralise their data input and output. For Unix/Linux based systems the *Network File System* (NFS) and for Windows systems the *Sever Message Block* (SMB) are the preferred choice. As the provision of remote access to a file system always increases the system's attack surface, it should by default be limited to those users that really have a need for it. Access rights should be specific on the detailed directory structure and the default right should be read-only, unless there is a need to write, or even execute. For Linux systems, there might also be additional hardening recommendations to be considered like the restriction of NFS to only use TCP and not UDP (refer to SLES 12 Hardening Guidelines [10]). Any remote file access service should by default be disabled on clients or servers that do not actually make use of it.

16.2.9   File Permissions

Every file on a computer system has a set of permission attached to it, which defines who can read, write, or execute it. In Unix/Linux systems, file permissions can be specified for the *user* (u), the *group* (g), and *others* (o), where the last category is also referred to as "everyone else" or the "world" group. Every operating system comes with a default file permission configuration which needs to be carefully revised and potentially adapted to the project's need. Special attention needs to be given to permissions defined for the "world" group. World-readable files should not

---

[6] A sandbox environment refers to a tightly controlled and constraint environment created for an application to run in. The aim of this is to limit the level of damage an application could cause in case it has been infiltrated with malicious code.

contain sensitive information like passwords, even if these are hashed and not human readable. World-writable files are a security risk since they can be modified by any user on the system. World-writable directories permit anyone to add or delete files which opens the door for the placement of malware code or binaries. A systematic search for such files and directories should therefore be performed on a regular basis and, if found, appropriate measures taken. A mitigation would be the addition of the so called *sticky bit* flag for the affected data item which is a specific setting in the properties that restricts the permission for its modification to the file owner or creator.[7] The permission to execute is a necessary condition to run an application or script on the system and should therefore be also carefully monitored, as it is also required by malicious code to operate.

16.2.10 User Account Management

User account management is another important security aspect, as it ultimately defines who can access a system, which areas can be accessed, and which application a user can execute. It is important that all user accounts on every machine are properly tracked, monitored, and maintained. Accounts that are not used anymore need to be identified, locked or deleted. It is also critical to track so called "non-human" accounts, which are accounts created for daemons, applications, or scripts that run on a system.

16.2.11 Password Policy

A password policy defines the rules that apply for the required complexity of a new password chosen by a user and any restrictions on the re-use of previously chosen ones. It also sets requirements on password renewal intervals, which can be enforced through the setting of password expiration time spans. The rules on password strength are usually defined by the project or subject to a company policy. Stricter rules will apply for systems that host sensitive data with some kind of security classification.

16.2.12 System Resource Limitation

Limiting the amount of system resources that users, groups, or even applications can consume is a method to avoid the slow down or even unavailability of a machine, in

---

[7] The name *sticky bit* is derived from the metaphoric view to make the item "stick" to the user.

case an application wants to use up all available system resources. Examples could be an uncontrolled build up of memory leaks, or the reservation of all available file handles. This can make a server unresponsive for any new user login or task, which is referred to as an accidental *Denial of Service* (DoS) event. There is also the intentional creation of a DoS event which is a widely established cyber attack vector and can cause major service interruptions for end users. The limitation of machine resources is therefore also an efficient means to counteract this type of DoS attacks.

16.2.13   Login Banners

Login banners refer to any information presented to a user when logging into a server from remote. This is also referred to as the *message of the day* (MOTD) and in the OS default configuration might presents sensitive system related information to the outside world. A typical examples is the type of OS and its version, which unintentionally provides an attacker important hints on potential zero-days existing on the target host. It is therefore recommended to always adapt the MOTD banner in order to avoid an unnecessary presentation of system related information and replace it with statements that deter intruders from illegal activities (e.g., a warning that any login and key strokes are being tracked).

16.2.14   System Integrity

The use of system integrity checkers are a powerful means to verify any unintentional or illegal modification on a system. A widely used tool is the *Advanced Intrusion Detection Environment* (AIDE) which is an open source package provided under the GNU General Public License (GPL) terms [11]. When initially run on a system, the application creates a database that contains information about the configuration state of the entire file system. This can be seen like taking a fingerprint or a snapshot of the system's current state. Whenever AIDE is rerun at a later stage, it compares the new fingerprint with the initial reference database and identifies any modification that has occurred since then. Potential unintended or forbidden configuration changes can then be easily spotted and further scrutinised.

16.3   Cyber Security Engineering

The aim of cyber security engineering is to ensure that during the design, development, and maintenance of an IT system all measures are taken to maximise its robustness against cyber attacks. Good practice cyber security engineering aims to consider relevant measures in the early design so they become part of the initial system architecture. This will not only improve the cyber robustness at first deployment, but also allow it to be kept up-to-date and robust throughout its entire operational lifespan. The upgrade of old infrastructure to improve the cyber resilience can be very challenging and can potentially require a major redesign activity. If this is the case, such a system upgrade requires careful planning in order to avoid any service interruption or if not otherwise possible reduce it to the minimum extent. Some generic principles for consideration are provided in the bullets below.

- Cyber security aspects must be taken into account as part of any system design decision that impacts the overall system architecture and must always aim to reduce the system's attack surface to the greatest possible extent. An example is the introduction of *air gaps* at highly sensitive system boundaries, which refers to the deliberate omission of a physical network connection between two devices for security purposes. This forces any data transfer to be performed via offline media which can be more easily tracked and controlled. Other examples are efficient and fast means to introduce new software patches to fix vulnerabilities throughout the entire system lifetime, the introduction of external and internal firewalls, the use of intrusion detection systems, the centralised management and control of user access, and efficient means to upgrade antivirus software and perform antivirus scanning, to name only a few.
- The regular planning and execution of obsolescence upgrades both at hardware and software level is an important means to keep the number of cyber vulnerabilities in a system low. Obsolete software will reach the end of vendor support period very soon which implies that no patches to fix vulnerabilities might be available anymore. The use of virtualisation technology (see Chap. 14) is highly recommended as it provides more flexibility in the modernisation process. If an already obsolete operating systems has still to be used for backwards compatibility with heritage software, adequate protective measures must be taken. An example would be the use of a virtual machine (VM) that is configured to have only very limited connectivity to the remaining system. For the absolutely necessary interfaces, additional protective measures should be deployed (e.g., firewalls, DMZ, etc.)
- The development and maintenance of *security operations procedures* (SECOPS) is an important building block to ensure and improve secure operations. Such procedures define and document a set of rules that reduce the risk for cyber attacks or sensitive data leakage and can be easily handed to system operators and maintainers as a guideline or even applicable document. Examples of topics defined in SECOPS procedures are the required time intervals to update antivirus (AV) definition files, schedules for the execution of AV scans, the enforcement of password complexity rules and change policies, the definition of user groups and rights, procedures for data export import and tracking, or port lock-down procedures.
- Penetration tests (also referred to as pentests) are designed and executed to unveil unknown vulnerabilities of an operational system and should always be followed

by a plan to fix any major discovery in a reasonable time frame. Good pentest reports should provide a summary of all discovered vulnerabilities, their location, how they could be exploited (risk assessment), and how they should be fixed. As all of this information is highly sensitive, extreme care must be taken with the distribution of such type of documentation, as it would provide a hacker a huge amount of information to attack and damage the entire system.[8]

16.4   Audit vs. Pentest

The cyber robustness of an IT system is never a stable situation and will by nature degrade over time due to the advent of new threats aiming to exploit new discovered vulnerabilities. It is therefore important to monitor and assess the cyber state of a system at regular intervals, which can be achieved by two different types of activities, the *cyber audit* and the *penetration test*.

The cyber audit aims to investigate the implementation of cyber security related requirements, both at design and operational level. At design level, such an audit could for example verify whether an air gap is actually implemented in the deployed system and corresponds to the system documentation. Another example is the inspection of the lock-down state of a system to verify whether the actual configuration (e.g., USB port access, network ports, firewall rules, etc.) is consistent with the required configuration. At operational level, the proper execution of applicable SECOPS procedures should be audited. This could comprise a check when the last antivirus scan was performed or whether the applicable password policy is actually being exercised and enforced.

The purpose of a penetration test, colloquially also referred to as *pentest*, is to unveil vulnerabilities of an IT system via a simulated cyber attack. The important difference to a real attack is that in the case of a pentest, the system owner is fully aware of it and has also agreed to it, in order to gain more knowledge. Every pentest needs to be carefully planned and executed and the detailed scope should not be known to the system owner in order to ensure a high degree of representativeness to a real case attack scenario. The outcome of a pentest must be a well documented and summarise all the findings together with a categorisation of their severity. The latter will give the owner an indication of the urgency to either fix a finding or to mitigate the potential damage that it could cause in the case of an attack. It is also useful to suggest potential countermeasures, both at short and long term level, which could reduce or even eliminate the risk of a vulnerability.

To effectively design and plan a pentest scenario, it is worth to apply the *Cyber Kill Chain* framework, which has been proposed by the Lockheed Martin Corporation [12, 13] and is well described in the open literature on computer

---

[8] The results of a pentest are usually subject to a security classification like restricted, confidential, or even secret, depending on the criticality of the system they have been collected from.

security (see e.g., [14]). The Kill Chain describes the various phases of the planning and execution of a cyber attack. It starts from early reconnaissance of the target system, followed by the selection of the appropriate malware ("weaponisation"), its delivery and installation onto the target system, and finally the remote control and execution of all kind of malicious actions (e.g., lateral expansion to other systems, data collection or corruption, collection of user credential, internal reconnaissance, destruction of the system, etc.). The Cyber Kill Chain framework also suggests countermeasures that can be taken during each of these phases in order to defend a computer network or system. These Kill Chain phases should be emulated by the pentest designer in order to make it more realistic and to better evaluate the effectiveness of defensive countermeasures. As a minimum, a pentest should focus on the following aspects:

- Intrusions from external and internal interfaces,
- inter-domain intrusions, e.g., between classified and unclassified parts of a system and even parts that are separated by air gaps,
- robustness against insider threats (e.g., possibilities to introduce malware),
- robustness against threats that could exploit all existing software vulnerabilities, and
- the functionality and efficiency of a system's integrity verification algorithm, in other words, to measure how quickly and reliably a potential contamination with malicious code can be discovered and appropriate measures taken.

16.5 Threat Analysis

A thorough and continuous analysis of the current cyber threat landscape must be a fundamental component in every efficient cyber security strategy. The *threat-driven approach* as described by Muckin and Fitsch in their White Paper (cf., [15]) resembles, from a methodology point of view, the classical fault analysis practices applied in systems engineering like for example the *failure mode effects analysis* (FMEA) and the *failure mode effects and criticality analysis* (FMECA). The respective authors advocate that threats should be the primary driver of a well designed and properly defended system, as opposed to approaches that simply comply to one or several pre-defined cyber security regulations, directives, policies or frameworks, which might not be fully adequate for a specific system architecture or already outdated. In other words, this approach places threats at the forefront of strategic, tactical and operational practice, which requires the incorporation of a thorough threat analysis and threat intelligence into the system development and operations. This enables the measures derived from such an approach to be tailored more specific to the system and its environment, also taking into account the context it is actually operated in. The authors propose a threat analysis method that defines a set of steps that can be grouped into a discovery and an implementation phase.

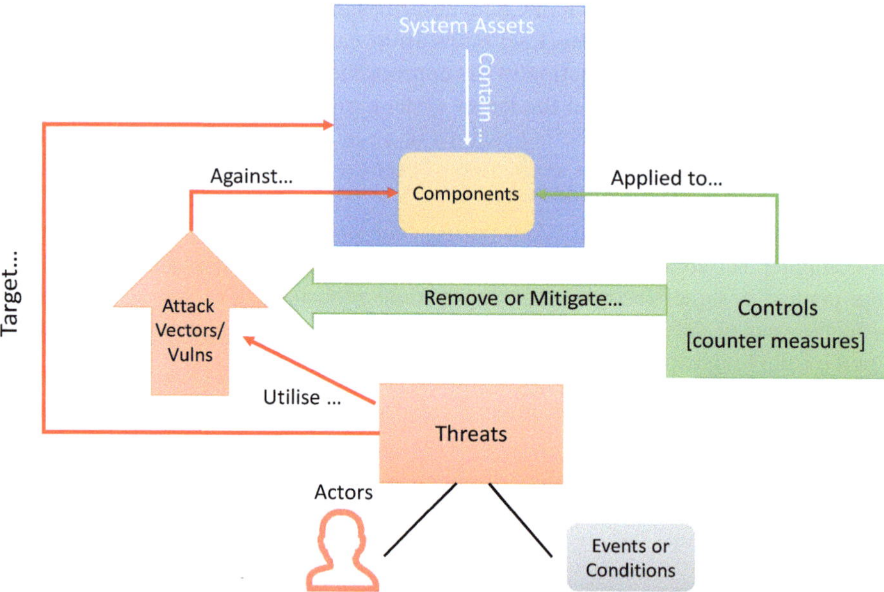

Fig. 16.2 Threats-Assets-Controls relationship model as defined by the threat-driven cyber security approach (based on Fig. 3 of [15])

These are briefly outlined below and the reader is referred to the cited literature for more details:

- The discovery phase focuses on the clear identification of the system's assets, which refers to the list of potential targets for a cyber attack (see also Fig. 16.2), followed by the determination of their attack surface. The identified assets are then decomposed into their technical components which could be devices, interfaces, libraries, protocols, functions, APIs, etc. This of course requires the detailed knowledge of the system and should be based on a thorough consultation of the relevant design documentation. The next step is to identify the list of potential attack vectors specific to the listed assets and their derived components. It is also recommended to analyse the threat actors, their objectives, motivation, and ways how the attack could be orchestrated as this might impact the list of assets identified in this step (this data is referred to as *threat intelligence*).
- The implementation phase first produces a prioritised listing (also referred to as "triage") considering business or mission objectives that contribute to the risk assessment of a threat. To give an example, a threat that might lead to the leakage of sensitive information (e.g., project design information, source code subject to corporate intellectual property, etc.) must be considered as a higher priority than threats affecting less sensitive data (e.g., a corporate directory). The final step in this activity is to implement the necessary *security controls* which are the counter measures that need to be taken in order to remove or at least mitigate the

identified threats and attack vectors. This is either done as part of the development or engineering work (in case the system is still under development), the roll-out of software patches, or the adaptation of configuration parameters in the system (e.g., lock down of interfaces or ports, etc.).

It should be noted that the threat analysis described above should be considered as one example of many and the reader is encouraged to also consult other approaches described in the following references: [16–18].

16.6 Cryptography

Cryptography is a field in computer science that aims to provide techniques that allow the exchange of a message between two parties and at the same time prevents a third party (referred to as adversary) to illegally intercept it. This basic idea of such a secure communication is depicted in Fig. 16.3 where the two parties A and B are nick-named "Alice" being the sender and "Bob" being the intended recipient. The adversaries are named "Eve" and "Mallory" and represent attackers that launch an eavesdropping or a man-in-the middle attack, respectively. The secure exchange of information between Alice and Bob is achieved through the use of a device that encrypts (or enciphers) an input plaintext p using an algorithmic function $E(p, k) = c$. The algorithm used for the encoding is referred to as a *cipher* and the resulting coded text c as ciphertext. The ciphertext text can now be sent through the insecure channel without any risk but it needs to be decrypted by the recipient using the invert function $D(c, k') = p$.

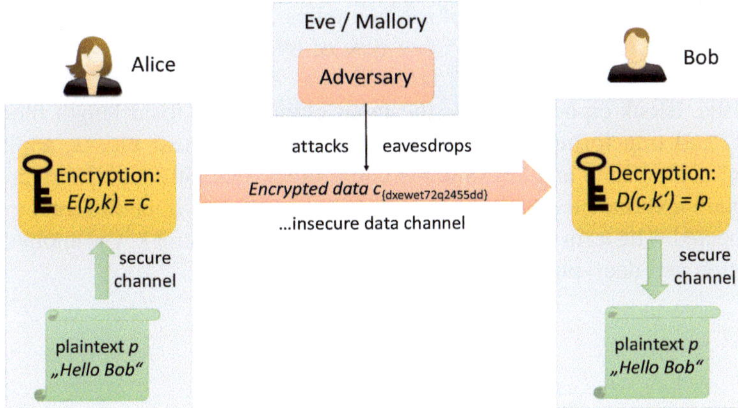

Fig. 16.3 Basic concept to establishing a secure communication channel using cryptography (refer to text for more detailed explanation)

As can be readily seen, both the encryption and decryption phase depend on the definition and knowledge of a key k that needs to be applied at both ends. The exchange and use of such keys is a non-trivial task as it ultimately defines the security of the entire system.[9] There are two main concepts in use for such an exchange, the *symmetric* and *asymmetric* key algorithms. The symmetric-key algorithms use the same cryptographic keys[10] for both the encryption of the plaintext and its decryption. Such keys represent a shared secret between the parties and therefore need to be exchanged on secure channels themselves. The asymmetric key concept is based on the generation of a pair of keys by each of the two parties involved. One of the two keys, the so called *public key*, can be openly distributed to another party who can then use it to encrypt a message. The decryption of that message will however only be possible using the second privately held key. This allows a secure transmission in one direction which can be used to exchange a symmetric key according to the *Diffie-Hellmann* public-key exchange protocol [21]. Another key exchange protocol was introduced a bit later and is referred to as the *Rivest-Shamir-Adleman* or RSA protocol [22]. It is also based on a public-key cryptosystem with different keys for encryption and decryption, but the key encryption process uses two large prime numbers. The key itself is public and available for anyone to encrypt a message but decryption is only possible for those who know the actual prime numbers.

Another important application of public-key cryptography is the digital signing of documents which allows to authenticate the sender's identity. This allows to demonstrate file integrity and to prove that the document has not been tampered during its transmission. In this case, the sender uses the private key for encryption (thereby signing it) and the recipient the public key for decryption (thereby verifying the signature).

The specification of an encryption methodology comprises the encryption algorithm, the key size, and the key symmetry. Key sizes are given in number of bits, keeping in mind that a key size of n-bits implies 2^n possible key combinations that will have to be tested in a brute-force type of attack. There are two basic types of symmetric encryption algorithms referred to as *block* and *stream* ciphers (refer to e.g., [19]). Block ciphers process the input plain text in fixed length blocks (e.g., 64 bits) and transforms these into corresponding ciphertext blocks, applying one encryption algorithm and key onto the entire block size. Stream ciphers process the input plain text one bit at a time, and that bit encryption will be different every time it is repeated. The term *symmetric* means that the same secret key is applied for both encryption and decryption.

---

[9] It is a fundamental concept of cryptography that all the secrecy must reside entirely in the key, as the cryptographic algorithm itself is known to the public (and an adversary). This is referred to as Kerckhoffs's assumption as it has been put forward by A. Kerckhoffs in the nineteenth century [19, 20].

[10] Or alternatively one can be computed from the other.

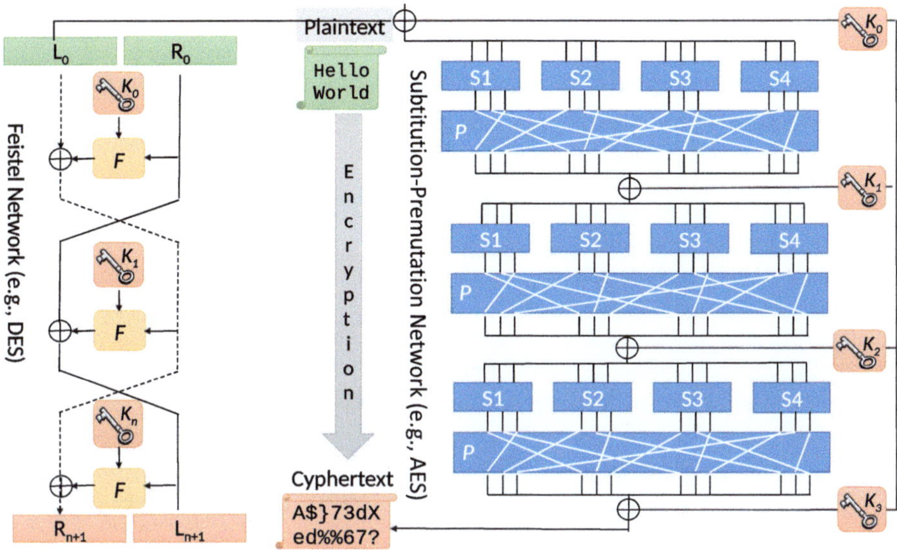

Fig. 16.4 Basic concept of block-ciphers (refer to text for detailed explanation). Left side: schema of a *Feistel* network as used in the Data Encryption Standard (DES); L_0 and R_0 = 32 bit half-blocks, F = *Feistel* function, $K_{i..n}$ = sub-keys used in round-i. Right side: schema of a substitution-permutation-network as used in the Advanced Encryption Standard (AES); $S_{1...n}$ = substitution boxes (S-boxes), P = permutation boxes

Two different block cipher techniques are depicted in Fig. 16.4. The left branch is used by an encryption method developed in the early seventies at IBM with contributions of the US National Security Agency (NSA). It has been published as an official Federal Information Processing Standard (FIPS) as *Data Encryption Standard* (DES) [23] and has been widely used as a standard encryption methodology in the past decades. The process starts by dividing a 64 bits plaintext block into two equal half-blocks of each 32 bits which are then alternatively fed to the encryption algorithm referred to as a *Feistel* function (F).[11] The F function performs several processing steps, starting with a bit expansion (i.e., from 32 bits to 48 bits), a sub-key ($K_{0..n}$) mixing stage, substitution, and permutation. In the last two steps the block is divided into so called substitution and permutation boxes, called S-, and P-Boxes respectively. During substitution, the input bits of the S-box are replaced by different output bits taken from a lookup table, and during permutation these S-boxes are rearranged. The output of F is then recombined with the corresponding half-block that has not been processed by F using an exclusive-OR (XOR) operation. This process is repeated several times (referred to

---

[11] Named after Horst Feistel, a German-born physicist and cryptographer who was one of its designers working for IBM [24].

as "rounds"), and in every round the block fed into the K-function is alternated. This criss-crossing process is known as a *Feistel-network* [25].

Due to its relatively short key size of only 56-bits, the DES encryption standard has been considered as insecure and was subsequently superseded by a more secure one, referred to as the *Advanced Encryption Standard* (AES) [26] shown on the right side of Fig. 16.4. AES uses longer keys (available lengths of 128, 192, and 256 bits) and a block size of 128 bits. It also performs several rounds with each one performing a substitution operation on a set of S-boxes. This is then followed by a permutation operation that takes the outputs of all the S-boxes of one round, permutes the bits, and feeds them into the S-boxes of the next round. Each round takes a different sub-key $K_{1...n}$ as input [25]. Due to its improved security, AES is today (2022) widely used in many internet protocols (e.g., transport layer security or TLS) and even for the handling of top secret information.

16.7   Concluding Remarks

The aim of this chapter was to provide the reader a basic introduction into the complex and very dynamic subject of cyber security. The threats stemming from cyber attacks are constantly growing, and so the potential damage they can cause. As both ground and space segment complexity is continuously growing and with it their presence in our strategically vital infrastructure, they have moved more and more into the focus of hackers. Therefore, cyber security has to be an integral part of a modern ground segment systems engineering process.

For a better assessment of a system's vulnerable areas, it is important to develop a good understanding of the various types of attack vectors, their dynamic evolution, and the potential damage they can cause. To support this, the concept of the attack surface has been introduced. Cyber security aspects need to be considered during the design, development, and operational phases of a new ground segment. For existing ones, obsolete infrastructure or software need to be updated or replaced as soon as possible. SECOPS procedures are an important contribution to identify the need for cyber patches and to enforce secure operational practices like system lock-downs, access control, or password complexity and renewal rules.

The need for cyber security audits and penetration tests has been explained, as both are of paramount importance to understand the current *cyber state* of a system. Their regular execution allows to unveil existing vulnerabilities and initiate their fixes, or to take measures to mitigate the risk for them to be exploited in a potential cyber attack. A basic understanding of the *Cyber Kill Chain* helps to "design" more efficient and representative penetration tests.

The concept of threat analysis has been introduced as a means to drive system design in relation to the specific cyber threat environment the system is exposed to. The described methodology is more effective compared to the simple application of a standard set of cyber requirements and procedures.

Basic concepts of cryptography were described with the aim to provide the reader a first insight into a mathematically complex subject matter. Unless the readers wants to become a cryptography expert, it is probably sufficient to understand the main features of cryptographic algorithms and protocols, which are strongly driven by the key size and the effort to generate and securely exchange them. It is important to understand the difference of existing encryption standards and the level of security they can offer. The need to apply up-to-date standards in early system design is obvious, but existing infrastructure might still use deprecated encryption protocols and ciphers which needs to be carefully analysed and upgraded, if possible. Encryption must be an integral part of any information flow in a modern ground segment. Special attention should be given to interfaces that connect to remote sites (e.g., TT&C stations) which usually make use of wide area network infrastructure that is not necessarily part or under control of the project specific infrastructure perimeter (e.g., rented telephone lines).

References

1. Knapp, E. D. (2011). Industrial network security: Securing critical infrastructure networks for smart grid, SCADA, and other industrial control systems. In *Syngress* (1st ed.). Syngress Publishing. ISBN-10: 9781597496452
2. Davidson, C. C., Andel, T. R., Yampolskiy, M., et al. (2018). On SCADA PLC and fieldbus cyber security. In *Proceedings of the 13th international conference on cyber warfare and security, National Defense University, Washington, DC*.
3. Wright, G., & Bacon, M. (2022). *Definition watering hole attack*. SearchSecurity. https://searchsecurity.techtarget.com/definition/watering-hole-attack, Accessed Apr 3, 2022.
4. European Union Agency for Cybersecurity (ENISA). (2020). From January 2019 to April 2020 Web-based attacks. ENISA Threat Landscape Report ETL-2020. https://www.enisa.europa.eu/publications Accessed Apr 2, 2022.
5. CVE List Home. (2022). https://cve.mitre.org/cve/, Accessed Apr 3, 2022.
6. CVE Search List. (2022). https://cve.mitre.org/cve/search_cve_list.html, Accessed Apr 3, 2022.
7. Manadhata, P. K., & Wing, J. M. (2011). A formal model for a system's attack surface. In *Advances in information security* (Vol. 54). Springer Science+Business media, LLC. https://doi.org/10.1007/978-1-4614-0977-9_1.
8. Lieberman Software Corporation. (2006). *White paper: Wake on LAN technology* (Rev 2). https://elielecatportefoliosio.files.wordpress.com/2016/04/wake_on_lan.pdf, Accessed Apr 4, 2022.
9. International Organization for Standardization. (1994). Information technology - open systems interconnection - basic reference model: The basic model. ISO Standard ISO/IEC 7498-1:1994. https://www.iso.org/standard/20269.html, Accessed Apr 4, 2022.
10. SUSE LLC. (2017). Security and hardening guide, SUSE linux enterprise server 12 SP3. https://www.suse.com/documentation, Accessed Apr 4, 2022.
11. Advanced Intrusion Detection Environment (AIDE). (2022). Github page https://aide.github.io/, Accessed Apr 4, 2022.
12. Lockheed Martin Corporation. (2015). Gaining the advantage, applying cyber kill chain methodology to network defense. https://www.lockheedmartin.com/content/dam/lockheed-martin/rms/documents/cyber/Gaining_the_Advantage_Cyber_Kill_Chain.pdf, Accessed Apr 4, 2022.

13. Lockheed Martin. (2015) Seven ways to apply the cyber kill chain with a threat intelligence platform. https://www.lockheedmartin.com/content/dam/lockheed-martin/rms/documents/cyber/Seven_Ways_to_Apply_the_Cyber_Kill_Chain_with_a_Threat_Intelligence_Platform.pdf, Accessed Apr 4, 2022.

14. Yadav, T, & Rao, A. M. (2015). Technical aspects of cyber kill chain. In J. Abawajy, S. Mukherjea, S. Thampi, & A. Ruiz-Martínez (Eds.), *Security in computing and communications (SSCC 2015)*. Communications in computer and information science (Vol. 536). Springer. https://doi.org/10.1007/978-3-319-22915-7_40.

15. Muckin, M., & Fitch, S. C. (2019). *A threat-driven approach to cyber security, methodologies, practices and tools to enable a functionality integrated cyber security organization*. Lockheed Martin Corporation, https://www.lockheedmartin.com/content/dam/lockheed-martin/rms/documents/cyber/LM-White-Paper-Threat-Driven-Approach.pdf, Accessed Apr 4, 2022.

16. Microsoft Corporation. (2022). *Microsoft security development lifecycle (SDL)*. https://www.microsoft.com/en-us/securityengineering/sdl/threatmodeling Accessed Apr 4, 2022.

17. Building Security In Maturity Model (BSIMM). (2022). http://www.bsimm.com/, Accessed Apr 4, 2022.

18. Synopsis. (2022). *Application security threat and risk assessment*. https://www.synopsys.com/software-integrity/software-security-services/software-architecture-design.html Accessed Apr 4, 2022.

19. Kahn, D. (1967). *The codebreakers: The story of secret writing*. Macmillan Publishing.

20. Schneier, B. (1996). *Applied cryptography: Protocols, algorithms and source code in C* (2nd ed.). John Wiley and Sons. ISBN 0-471-11709-9.

21. Merkle, R. C. (1978). Secure communications over insecure channels. *Communications of the ACM, 21*(4), 294–299. CiteSeerX 10.1.1.364.5157. https://doi.org/10.1145/359460.359473. S2CID6967714.

22. Rivest, R., Shamir, R. A., & Adleman, L. (1978). A method for obtaining digital signatures and public-key cryptosystems. *Communications of the ACM, 21*(2), 120–126. CiteSeerX 10.1.1.607.2677. https://doi.org/10.1145/359340.359342.S2CID2873616.

23. Federal Information Processing Standard FIPS. (1999). *Data encryption standard (DES)*. National Institute of Standards and Technology (NIST), Publication 46-3.

24. Feistel, H. (1973). Cryptography and computer privacy. *Scientific American, 228*(5), 15–23.

25. Stinson, D. R., & Paterson, M. B. (2019). *Cryptography theory and practice* (4th ed.). CRC Press Taylor & Francis Group.

26. Federal Information Processing Standard FIPS. (2001). *Advanced encryption standard (AES)*. National Institute of Standards and Technology (NIST). Publication 197.

Appendix A
Coordinate Systems

The definition of coordinate systems is a very important aspect in every space project, as it enables the correct exchange of important data between the launch, space, and ground segment. It is also highly relevant for the correct transfer of orbit information to external entities, which could be a rented TT&C antenna from an external station network, an external ground segment supporting a specific phase of the project (e.g., LEOP or IOT), or another space agency that is part of an international cooperation. The various coordinate systems used in a project should therefore be described in a formal project document. Even if some of them will usually be very standard ones and well described in astrodynamics literature, others might be very project specific and depend on the orbit geometry, specific features of the satellite platform, or the type and mounting of sensors in use. It is therefore important to develop a basic understanding of the most commonly used coordinate systems, and a brief overview is presented here. For a more detailed explanation, the reader is referred to more specialised literature on astrodynamics (e.g., [1, 2]).

One fundamental aspect of a coordinate system is whether it can be considered as *inertial* or *non-inertial*, as this impacts the formulation of an object's dynamics in it. For an inertial system, the origin and axes can be considered as not accelerated with respect to inertial space (celestial frame). This makes the formulation of the satellite's equations of motion simple, as Newton's first law of motion holds and the second law can simply be expressed in the form $\vec{F} = m\,\vec{a}$.

For a non-inertial frame however,[1] the equations of motion need to consider so called *fictitious* forces. Examples are the Coriolis force, the centrifugal force, or the Euler force, which should not be interpreted as real physical forces, but terms that account for the movement of the coordinate system itself. A typical example for a non-inertial frame is the body-fixed (rotating) system, with one axis aligned to the angular velocity vector of a planetary body and its rotation synchronised to it. This makes the coordinate system to appear fixed with respect to the body's surface

---

[1] E.g., a frame that is subject to a rotation.

© The Author(s), under exclusive license to Springer Nature Switzerland AG 2023
B. Nejad, *Introduction to Satellite Ground Segment Systems Engineering*, Space
Technology Library 41, https://doi.org/10.1007/978-3-031-15900-8

(terrestrial frame) and allows a convenient way to specify a set of fixed coordinates
(e.g., longitude, latitude, and height) for any point on the surface.

Another group of coordinate systems are connected to the satellite body and
its orbital movement. The orbital system (LVLH) is a frequently used example to
express a satellite's thrust directions in radial, tangential, and normal components.
The satellite body frame and the instrument frame are used to specify geometric
aspects of the satellite's mass properties (centre of mass location, inertia tensor), the
location and orientation of an instrument sensor (mounting matrices), and the sensor
field-of-view.

A.1   Celestial Systems

A celestial reference systems is a systems who's centre coincides and moves with
the centre of a planetary body but does not rotate with it. As it is a system free of
rotational motion, it can be considered as an inertial or Newtonian reference system.
An example is the *Earth Centred Inertial* or ECI system which has its origin at the
Earth's centre (see Fig. A.1) and the z-axis aligned with the Earth's rotation axis
and pointing to the north pole. The $x-$ and y-axes both lie in the equatorial plane,
where x points to the *vernal equinox*,[2] and the y-axis completes an orthogonal right-
handed system.

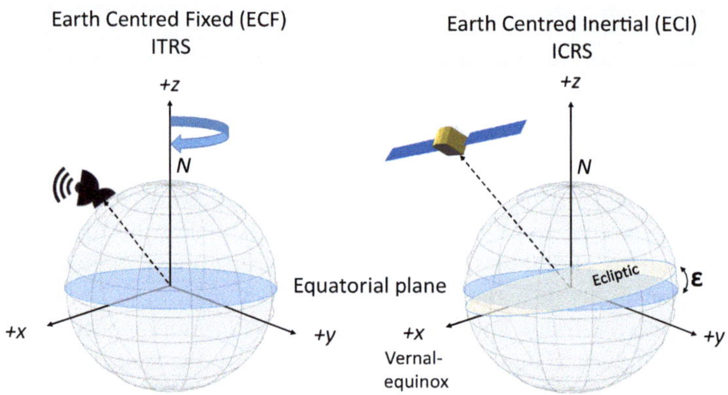

Fig. A.1 Definition of the *Earth Centred Fixed* (left) and the *Earth Centred Inertial* (right)
coordinate systems. The obliquity angle $\epsilon \approx 23.5°$ defines the inclination between the ecliptic
and equatorial planes whose intersection defines the location of the autumnal and vernal equinoxes

---

[2] One of the two intersection points of the equatorial plane with the ecliptic plane.

If the Earth's motion around the Sun would only be determined by the central force of the Sun, its plane would remain fixed in space. However, perturbations stemming from the other planets in the Solar system and the non-spherical shape of the Earth (equatorial bulge) add a torque on the rotation axis and cause a secular variation known as planetary precession that resembles a gyroscopic motion with a period of about 26,000 years [1]. As a result, the vernal equinox recedes slowly on the ecliptic, whereas the angle between the equatorial and the ecliptic plane (obliquity ϵ) remains essentially constant. In addition to the gyroscopic precession, short term perturbations of about one month can be observed which is called *nutation*. In view of the time dependent orientation of the Earth's equator and ecliptic, the standard ECI frame is usually based on the mean equator and equinox of the year 2000 and is therefore named *Earth Mean Equator and Equinox of J2000* or simply *EME2000*.

A.2   Terrestrial Systems

A terrestrial coordinate system has its origin placed in the centre of a rotating body. In contrast to the inertial (ECI) system, the $x-$ and $y-$ axes rotate with the bodies' angular velocity. The main advantage of a body fixed system is that any point defined on the body's surface can be conveniently expressed by constant coordinate components (e.g., longitude, latitude, and height above a reference surface). For the Earth the *International Terrestrial Reference System* or ITRS[3] provides the conceptual definition of a body-fixed reference system [5]. Its origin is located at the Earth's centre of mass (including oceans and atmosphere), and its $z-$ axis is oriented towards the *International Reference Pole* (IRP) as defined by the International Earth Rotation Service (IERS). The time evolution of the ITRS is defined to not show any net rotation with respect to the Earth's crust.

For the transformation of a position vector of a satellite in orbit expressed in ECI/EME2000 coordinates ($\vec{r}_{ECI}$) to its corresponding position on the Earth's surface ($\vec{r}_{ITRS}$), a coordinate transformation between the ECI and ITRS system has to be performed. This transformation is quite involved as it needs to take into account several models, i.e.,

- the precession or the Earth's rotation axis,
- the nutation describing the short-term variation of the equator and vernal equinox,

---

[3] Taking formal terminology strictly, the term reference *system* is used for the theoretical definition of a system, which comprises the detailed description of the overall concept and associated models involved. The term reference *frame* refers to a specific realisation of it, which is usually based on some filtering of measured coordinates from various ground stations.

- the *Greenwich Mean Siderial Time* (GMST), describing the angle between the mean vernal equinox and the Greenwich Meridian at the time of coordinate transformation, and
- the Earth's polar motion describing the motion of the rotation axis with respect to the surface of the Earth.

Taking all these components into account, the transformation can be expressed by a series of consecutive rotations in the following order[4]

$$\vec{r}_{ECI} = \Pi(t)\,\Theta(t)\,\mathbf{N}(t)\,\mathbf{P}(t)\,\vec{r}_{ITRS} \tag{A.1}$$

where $\mathbf{N}$ and $\mathbf{P}$ refer to the rotation matrices that describe the coordinate changes due to nutation and precession, respectively. The matrix Θ considers the Earth rotation and can be expressed as

$$\Theta(t) = \mathbf{R}_z(GAST) \tag{A.2}$$

where $\mathbf{R}_z$ is the rotation axis around the $z-$axis, and $GAST$ is the Greenwich apparent sidereal time which is given by the *equation of the equinoxes*

$$GAST = GMST + \Delta\psi \cos\epsilon \tag{A.3}$$

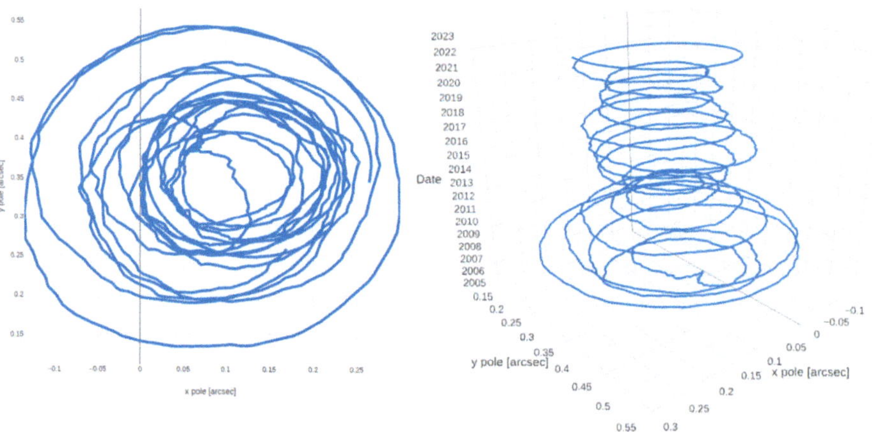

Fig. A.2 Polar motion components x_p and y_p of the Celestial Intermediate Pole expressed in the ITRS as published by the International Earth Rotation Service (IERS) [3]. Data retrieved from IERS online data centre [4]

---

[4] For a more detailed description of this involved transformation the reader is referred to specialised astrodynamics literature (e.g., [1, 2]).

where ψ is the angle between true and mean equinox and $GMST$ the Greenwich mean sidereal time. The Polar motion rotation matrix Π is given by

$$\Pi = \mathbf{R}_y(-x_p)\,\mathbf{R}_x(-y_p) \tag{A.4}$$

where x_p and y_p refer to the angular coordinates that describe the motion of the Celestial Intermediate Pole (CIP) in the International Terrestrial Reference System (ITRS) which is published in the Bulletins A and B of the IERS (cf., [3] and see Fig. A.2 for example data).

A.3   The Orbital Frame

The orbital coordinate frame is a right handed coordinate system with its origin located in the satellite's centre of mass and moving with its orbital motion. It is therefore convenient to define thrust orientation vectors for orbit correction manoeuvres. The unit vectors of its coordinate axes can be derived at any time from the satellite's inertial position and velocity vectors $\vec{r}(t)$ and $\vec{v}(t)$ defined in the ECI system, using the following simple relations

$$\vec{e}_R = -\frac{\vec{r}}{|\vec{r}|}$$

$$\vec{e}_N = -\frac{\vec{r} \times \vec{v}}{|\vec{r} \times \vec{v}|}$$

$$\vec{e}_T = \vec{e}_N \times \vec{e}_R \tag{A.5}$$

where the subscripts R, N, and T refer to the radial, normal, and tangential directions, respectively. As shown in Fig. A.3, the radial axis points to the central body (nadir direction), the normal axis is anti-parallel to the orbit angular momentum vector, and the tangential axis completes the right-handed system (and is close to the satellite's inertial velocity vector). The axis orientation shown here is also referred to as the *Local Vertical Local Horizontal* or LVLH-frame. An alternative definition of the orbital system is also in use, which has the radial direction defined as pointing away from the central body and the normal direction parallel to the orbital angular momentum vector.

A.4   The Satellite Body Frame

The satellite body frame is an important coordinate system that is required for the definition of the exact location of any device or sensor mounted on the satellite structure. Its origin and exact orientation have to be defined and documented by

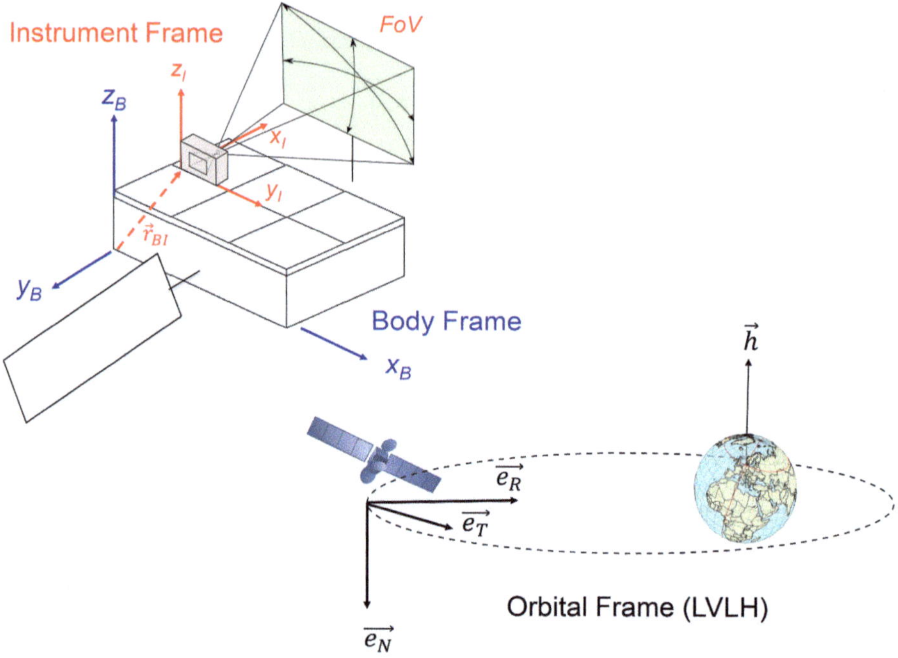

Fig. A.3 Definition of the satellite body frame (x_B, y_B, and z_B), the instrument frame (x_I, y_I, and z_I), and the orbital or Local Vertical Local Horizontal (LVLH) frame ($\vec{e}_R$, $\vec{e}_N$, and $\vec{e}_T$ referring to the radial, normal, and tangential unit vectors

the satellite manufacturer. The orientation will usually follow some basic alignment of the satellite structural frame as shown by the axes labelled as x_B, y_B, and z_B in Fig. A.3. The body frame is important for the definition of geometry dependant mass properties, like the moments of inertia tensor or the position of the centre of gravity and its movement due to change of propellant (refer to Chap. 3).

A.5 The Instrument Frame

The instrument frame is used to define the field-of-view (FoV) of a payload sensor mounted on the satellite structure. The ability to express the orientation of a sensor's FoV is important for the prediction of sensor related events, like the transition of a celestial body (e.g., the time the Sun enters the FoV of a Sun Sensor). The instrument frame origin is a reference point defined with respect to the satellite body frame and defines the position of the sensor on the satellite (see vector $\vec{r}_{BI}$ in Fig. A.3). The orientation of the axes x_I, y_I, and z_I is given by the so called *instrument mounting matrix* which is a direction-cosine matrix that defines the transformation between the two systems.

References

1. Montenbruck, O., & Gill, E. (2000). *Satellite orbits* (1st ed.). Springer Verlag.
2. Vallado, D. A. (2001). *Fundamentals of astrodynamics and applications* (2nd ed.). Space Technology Library, Kluwer Academic Press.
3. International Earth Rotation Service. (2014). *Explanatory supplement to IERS bulletin A and bulletin B/C04*. https://www.hpiers.obspm.fr/iers/bul/bulb_new/bulletinb.pdf, Accessed Apr 4, 2022.
4. International Earth Rotation and Reference Systems. (2022) Service website: https://www.iers.org, Accessed Apr 4, 2022.
5. McCarthy, D. D. (1996). *IERS conventions (1996)*. IERS Technical Note 21, Central Bureau of IERS, Observatoire de Paris.

Appendix B
Time Systems

The definition of time systems is a very complex subject matter and can therefore not be addressed here with the level of detail that it might deserve, and only some basic explanations are provided.[1] A basic understanding of the most common time systems is important, as every computer system in a ground segment must be synchronised to a single time source and system. Furthermore, any product exchanged with an external entity (e.g., telemetry, orbit files, or command parameters) must be converted to the time system that is defined in the applicable interface control document. It could even be the case, that a ground segment interfaces to various external entities and each of them works in a different system and expects the products to be converted accordingly.

Prior to the invention of atomic clocks, the measurement of time was mainly based on the length of a day, being defined by the apparent motion of the Sun in the sky. This gave the basis for the *solar time* which is loosely defined by successive transits of the Sun over the Greenwich meridian (i.e., the 0° longitude point). Due to the Earth's annual orbital motion, it rotates slightly more than 360° during one solar day. To avoid this inconvenience, the *siderial time* was defined as the time between successive transits of the stars over a particular meridian.

Unfortunately, things are even more complex. Due to the elliptical shape of the Earth's orbit around the Sun, it moves at variable speed. In addition, the obliquity between the celestial equator and the ecliptic gives raise to an apparent sinusoidal motion around the equator, which adds even more irregularities that are inconvenient for the measurement of time. Therefore, the *mean solar time* was introduced and based on a fictitious *mean Sun* having a nearly uniform motion along the celestial equator. The mean solar time at Greenwich is defined as *Universal Time* (UT) which has three distinct realisations, UT0, UT1, and UT2. UT0 is derived (reduced) from observations of stars from many ground stations. Adding corrections for

---

[1] For a deeper treatment, the reader is referred to more specialised literature (e.g., [1]).

B. Nejad, *Introduction to Satellite Ground Segment Systems Engineering*, Space Technology Library 41, https://doi.org/10.1007/978-3-031-15900-8

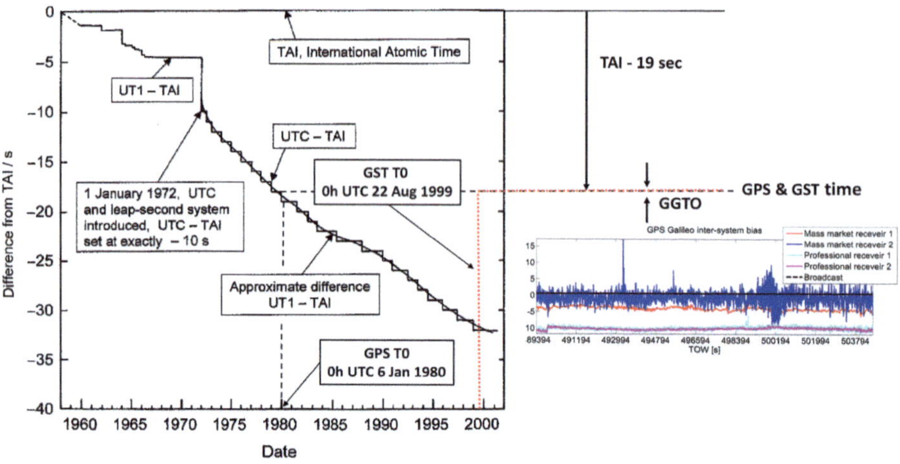

Fig. B.1 Difference between TAI, UT1, and UTC for the past decades as depicted in Nelson et al. (2003) [2]. The start epoch of the GNSS based GPS and GST time systems are shown as "GPS T0" and "GST T0". Both systems have a constant offset to TAI of 19.0 seconds. The difference between GPS and GST time (GGTO) is broadcasted as part of the Galileo Open Service (OS) navigation message [3]. Examples for the broadcast GGTO values are shown in the right panel (mean value ca. 0.59 m or $\approx$ 2 ns). Also shown are ground receiver measured GGTO values with a range between 0.65 m to 10 m ($\approx$ 2–33 ns) [4]

location dependant polar motion provides UT1,[2] and further corrections for seasonal variations provides UT2.[3]

A completely different source for a highly accurate and stable measurement of time came with the invention of atomic clocks, which use the well known low-energy (hyperfine) state transitions of specific atoms or molecules (e.g., Cesium, Hydrogen, or Rubidium). Using microwave resonators that are able to excite these states, these can be accurately tuned to achieve a maximum population density of excited atoms. This can be measured, and is used to keep the microwave resonator frequency at a highly stable reference value. In 1972, the atomic time scale was established at the french *Bureau International des Poids et Mesures* (BIPM), and adopted as a standard time under the name *International Atomic Time*. TAI is physically realised with an elaborate algorithm that processes readings from a large number of atomic clocks located in various national laboratories around the world (cf., [5]). A different atomic time scale is realised by GNSS satellite constellations like GPS or Galileo, which provide respectively the GPS time (GPST) and the Galileo System Time (GST). Both time systems are kept at a constant offset of exactly 19 seconds to TAI (see Fig. B.1).

---

[2] UT1 is therefore the same around the world and does not depend on the observatory's location.

[3] UT2 is not really used anymore and can be considered obsolete.

The most commonly used time system is *Coordinated Universal Time* or UTC which differs by an integer number of seconds from TAI and forms the basis for civilian time keeping. UTC is also used to define the local time around the world, by adding regionally dependant offsets as defined by the time zone and the use of daylight saving time (DST). UTC is derived from atomic time but periodically adjusted by the introduction of *leap seconds* at the end of June and/or the end of December in order to keep the difference $\Delta UT1 = UT1 - UTC \leq \pm 0.9^s$ (see Fig. B.1). The adjustment of UTC via the introduction of leap seconds is needed to compensate for the gradual slow down of the Earth's rotation rate, which is furthermore affected by random and periodic fluctuations that change the length of day relative to the standard reference day of exactly 86,400 SI seconds. It should be noted that the motivation to continue the introduction of new leap seconds has diminished in the past years due to the availability of satellite navigation time scales and the operational complexity to incorporate leap seconds on a regular basis. It is even debated to discontinue the introduction of new leap seconds entirely, meaning that $|UTC - UT1|$ could exceed 0.9 seconds in the future. A final decision has so far (2022) not been taken, but the topic continuous to be discussed in various working groups of international scientific organisations [2].

As UTC might be used for both time keeping and the communication with external entities, the design of the ground control segment needs consider the ability to introduce new leap seconds whenever published and ensure the correct number is considered for time conversions.

References

1. Vallado, D. A. (2001). *Fundamentals of astrodynamics and applications* (2nd ed.). Space Technology Library, Kluwer Academic Press.
2. Nelson, R. A., McCarthy, D. D., et al. (2003). The leap second: Its history and possible future. *Metrologia, 38*(6), 509–529. https://doi.org/10.1088/0026-1394/38/6/6.
3. Hahn, J., & Powers, E. (2005). Implementation of the GPS to Galileo time offset (GGTO). In *Joint IEEE International Frequency Symposium and Precise Time and Time Interval (PTTI) Systems and Applications Meeting, Vancouver.* https://doi.org/10.1109/FREQ.2005.1573899.
4. Gioia, C., & Borio, D. (2015). GGTO: Stability and impact on multi-constellation positioning. In *ION GNSS* (pp. 245–249). Institute of Navigation, JRC94736.
5. Guinot, B. (1989). Atomic time. In J. Kovalevsky, I. I. Mueller, & B. Lolaczek (Eds.), *Reference frames in astronomoy and geophysics. Astronomy and space science library* (Vol. 154, pp. 379–415). Kluwer Academic Publishers.

Acronyms

ACS	Attitude Control System
ACU	Antenna Control Unit
AES	Advanced Encryption Standard
AIT	Assembly, Integration, and Test
AIV	Assembly, Integration, and Verification
AND	Alphanumeric Display
AOCS	Attitude and Orbit Control System
AOS	Acquisition of Signal
AR	Acceptance Review
ASW	Avionic Software
AV	Anti Virus
BBM	Baseband Modem
BCR	Battery Charge Regulator
BDR	Battery Discharge Regulator
BER	Bit Error Rate
BITE	Built-in Test
BMU	Battery Monitoring Unit
BWFN	Beamwidth-between-first-nulls
CADU	Channel Access Data Unit
CAN	Controller Area Network
CCSDS	Consultative Committee for Space Data Systems
CCR	Configuration Change Request
CDR	Critical Design Review
CFI	Customer Furnished Item
CI	Configuration Item
CISC	Complex Instruction Set
CLTU	Command Link Transfer Unit
CMM	Carrier Modulation Mode
CMS	Configuration Management System

CPU	Central Processing Unit
CRR	Commissioning Result Review
CoM	Centre of Mass
COTS	Commercial off the Shelf
CS	Contact Schedule
CVE	Common Vulnerabilities and Exposures
DAL	Design Assurance Level
D/C or DC	Down Converter
DG	Diesel Generator
DES	Data Encryption Standard
DDoS	Distributed Denial of Service Cyber attack
DR	Decommissioning Review
DRR	Disposal Readiness Review
DTR	Digital Tracking Receiver
ECSS	European Cooperation for Space Standardization
EIRP	Effective Isotropic Radiated Power
ELR	End-of-Life Review
EM	Engineering Module
EMC	Electromagnetic Compatibility
EGSE	Electronic Ground Support Equipment
ER	Equipment Room
ERT	Earth Received Time
ESA	European Space Agency
ESD	Electrostatic Discharge
FBSE	Functions-Based System Engineering
FCT	Flight Control Team
FD	Flight Director
FDDB	Flight Dynamics Database
FDF	Flight Dynamics Facility
FDIR	Failure Detection Isolation and Recovery
FM	Flight Module
FOP	Flight Operations Procedure
FoV	Field of View
FRR	Flight Readiness Review
GCS	Ground Control Segment
GEO	Geostationary Orbit
GNSS	Global Navigation Satellite System
GOM	Ground Operations Manager
GOP	Ground Operations Procedures
GPST	GPS Time
HCI	Hyperconverged Infrastructure
HK	Housekeeping
H/O	Hand-Over
HPA	High Power Amplifier
HPBW	Half-power-beamwidth

HPC	High Priority Command
HPTM	High Priority Telemetry
H/W	Hardware
ICD	Interface Control Document
IERS	International Earth Rotation and Reference System Service
IF	Intermediate Frequency
IID	Incremental and Iterative Development
IMU	Inertial Measurement Unit
INCOSE	International Council of Systems Engineering
IT	Information Technology
KMF	Key Management Facility
KPA	Klyston Power Amplifier
LCC	Life Cycle Cost
LDAP	Lightweight Directory Access Protocol
LEO	Low Earth Orbit
LEOP	Launch and Early Orbit Phase
LNA	Low Noise Amplifier
LOS	Loss of Signal
LRR	Launch Readiness Review
LS	Launch Segment
LSP	Launch Service Provider
LTP	Long Term Plan
MBSE	Model Based Systems Engineering Approach
M&C	Monitoring and Control
MCF	Mission Control Facility
MCID	Master Channel ID (CCSDS)
MCR	Mission Concept Review (NASA milestone)
MCR	Mission Closeout Review (ESA/ECSS milestone)
MEO	Medium Earth Orbit
MIB	Mission Information Base
MIU	Modem Interface Unit
MLI	Multi-layer insulation
MITM	Man-in-the-Middle Cyber attack
MMFU	Mass Memory and Formatting Unit
MMI	Man Machine Interface
MOCD	Mission Operations Concept Document
MoI	Moments of Inertia matrix
MP	Mission Plan
MPF	Mission Planning Facility
MPP	Mission Planning Process
MTP	Mid Term Plan
MW	Momentum Wheel
NASA	National Aeronautics and Space Administration
NFS	Network File System
NCR	Non-Conformance-Report

NORAD	North American Aerospace Defense Command
NSA	National Security Agency
NTP	Network Time Protocol
OBC	Onboard Computer
OBS	Organisational Breakdown Structure
OBSW	Onboard Software (similar to ASW)
OBT	Onboard Time
OCM	Orbit Control Mode
ODB	Operational Database
OMG	Object Management Group
OOP	Onboard Orbit Propagator
OPF	Operations Preparations Facility
ORR	Operational Readiness Review
OS	Operating System
OVP	Operational Validation Plan
PCDU	Power Control and Distribution Unit
PDR	Preliminary Design Review
PDU	Power Distribution Unit
PIM	Platform Independent Model
PLOP	Physical Layer Operations Procedure
PO	Project Office
POAR	Planning and Offline Analysis Room
PR	Planning Request
PSM	Platform Specific Model
PSR	Pre-Shipment Review
QA	Quality Assurance
QR	Qualification Review
RF	Radio Frequency
RISC	Reduced Instruction Set
RW	Reaction Wheel
SADM	Solar Array Drive Mechanism
SAFe®	Scaled Agile Framework
SAR	System Acceptance Review
SATMAN	Satellite Manufacturer
SCF	Satellite Control Facility
SCID	SCID = Spacecraft ID (CCSDS)
SCTC	System Compatibility Test Case
SDM	Shunt Dump Module
SDR	System Definition Review
SE	Systems Engineering
SECOPS	Security Operations (procedures)
SIM	Simulator
SIR	System Integration Review
SLE	Space Link Extension
SM	Safe Mode

SMB	Server Message Block
SNR	Signal-to-Noise ratio
SMDL	Simulation Model Definition Language
SMP	Simulation Model Portability
SNMP	Simple Network Management Protocol
SOM	Spacecraft Operations Manager
SOR	Special Operations Room
SOW	Statement of Work
SPA	Solar Power Assembly
SPACON	Spacecraft Controller
SPR	Software Problem Report
SRD	Software Requirements Document
SRDB	Spacecraft Reference Database
SRR	System Requirements Review
SS	Space Segment
SSPA	Solid State Power Amplifier
SSR	Solid State Recorder
SSRR	Space Segment Reference Architecture
STP	Short Term Plan
SU	Software Unit
SVT	System Verification Test
SysML	System Modelling Language
TAI	Temps Atomique International (International Atomic Time)
TC	Telecommand
TCXO	Temperature Controlled Oscillator
TDD	Test Driven Development
TFVN	Transfer Frame Version Number (CCSDS)
TM	Telemetry
TMM	Thermal Mathematical Model
TP	Test Procedure
TPF	Task Parameter File
TRR	Test Readiness Review
TRB	Test Review Board
TSP	Time Source Provider
TT&C	Tracking, Telemetry, and Command
TWTA	Travelling Wave Tube Amplifier
UML	Unified Modelling Language
UPS	Uninterruptible Power Supply
U/C or UC	Up-Converter
UT	Universal Time
UTC	Coordinated Universal Time
VCB	Verification Control Board
VCD	Verification Control Document
VM	Virtual Machine
VPN	Virtual Private Network

VSM	Virtual System Model
WAN	Wide Area Network
WBS	Work Breakdown Structure
WP	Work Package
XML	Extensible Markup Language